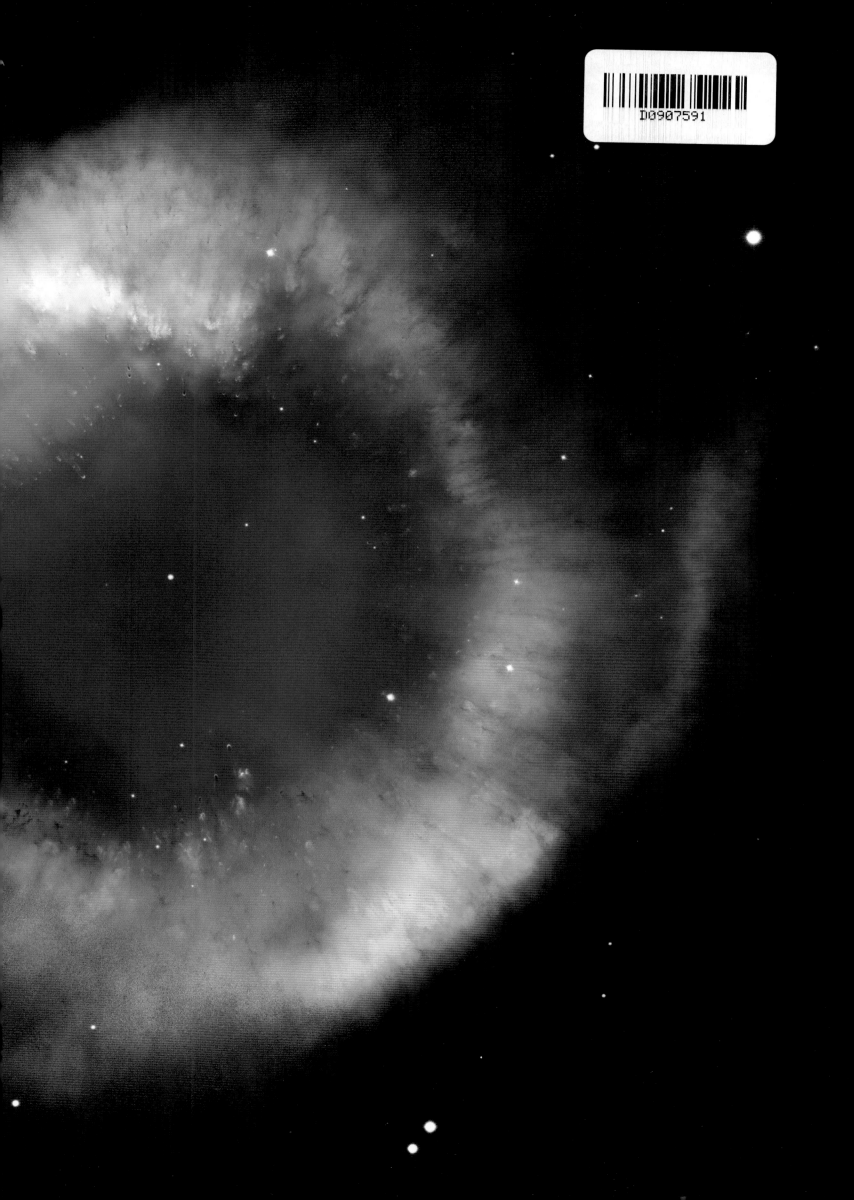

An exploration, in words and
pictures, of the thrilling history and inspiring
future of space travel and discovery

COMPANION TO THE TOURING EXHIBIT

TEXT BY ROGER D. LAUNIUS

a JOU U

SPACE
A JOURNEY TO OUR FUTURE

FU

When a star is in its death throes, previous spread, it puts on a spectacular final show. This image of the Helix Nebula shows a fine web of "bicycle-spoke" features in a red and blue ring of glowing gas more than 650 light-years away around a dying, star similar to our own Sun. The Helix Nebula is one of the nearest planetary nebulae to Earth. Overleaf: One of the most photogenic of galaxies, Messier 104 (M104), better known as the Sombrero galaxy, has a brilliant white, bulbous core encircled by thick dust lanes that give it the appearance of a high-topped Mexican hat. The most prominent of the objects in the Virgo cluster of galaxies and as bright as 800 billion suns, M104 is fifty thousand light-years across and is located 28 million light-years from Earth.

SPACE

RNEY

to OUR

TURE

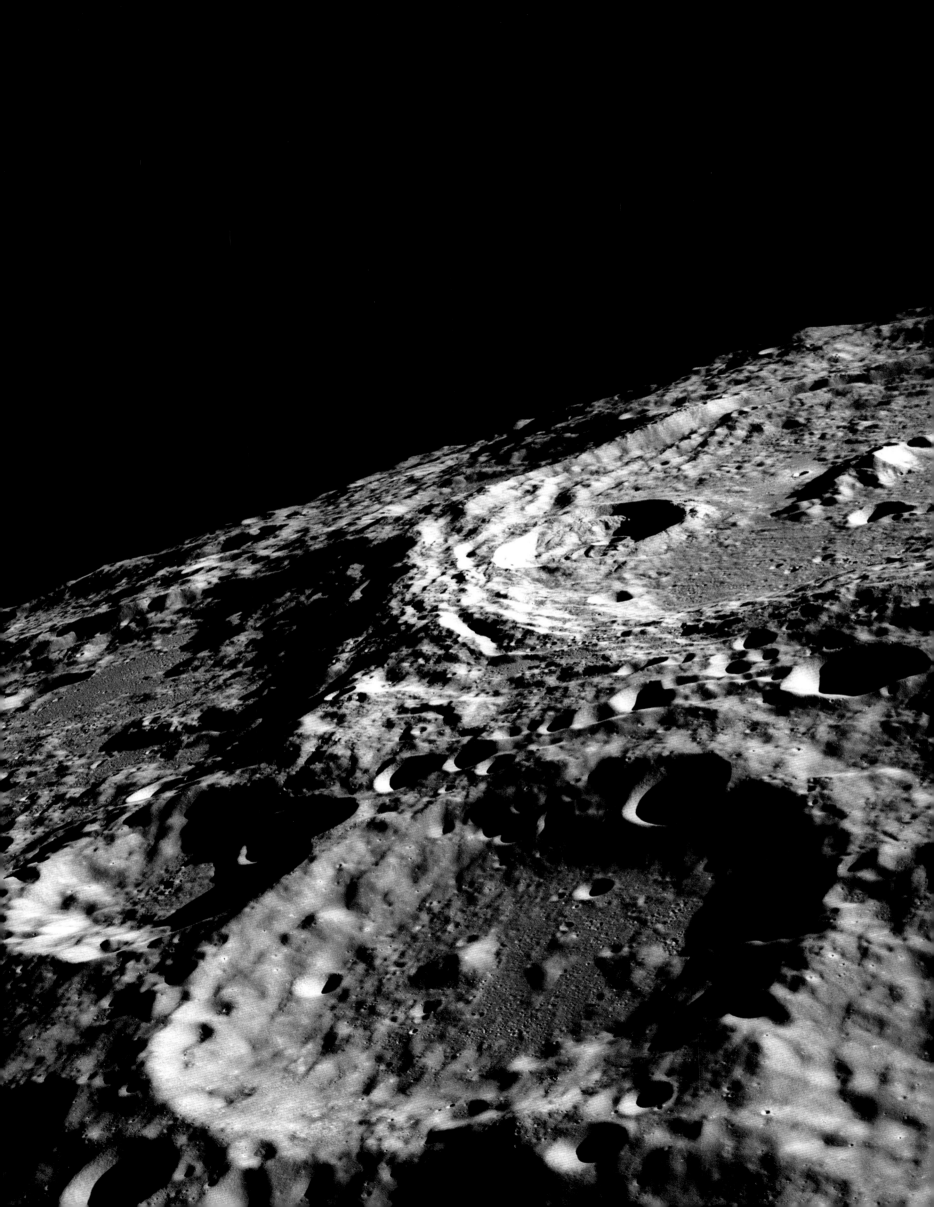

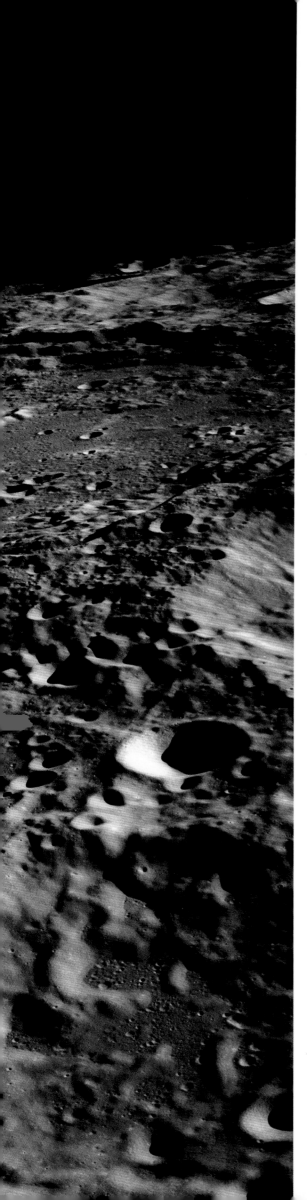

CREATING A JOURNEY FOR ALL

The crew of Apollo 10 captured the cratered surface of the Moon with a handheld 70mm camera on May 1, 1969. This view of the lunar surface features Crater 302 in the center foreground. If you look closely at the rim, you can see evidence of landslides caused by slow erosion from meteoritic impact, moonquakes, and heating-and-cooling expansion. Buzz Aldrin aptly referred to this lunarscape as "magnificent desolation."

W E AT CLEAR CHANNEL EXHIBITIONS are proud to present *Space: A Journey to Our Future,* a spectacular, multi-sensory exhibit that is the culmination of a personal passion for those of us who have worked for almost a decade to bring it to you.

In conceiving *Space,* we sought to develop an educational and entertaining experience that would immerse families in the thrilling story of the past, present, and — most importantly — future of space travel. As with all our exhibits, we tried to develop an experience that educates and inspires in the most exciting and engrossing ways possible.

In this exhibit, the theme of exploration runs deep. Truly, the urge to explore is innate to all human beings, and space exploration is a grand expression of that drive. We have aimed to give exhibit-goers the ability to delve into the subject of space travel themselves, and in the course of their journey, learn more about the cosmos.

It is our belief that thrilling educational experiences like this one may well inspire the future of space exploration: Today's young people are indeed tomorrow's scientists, engineers, and astronauts, and encouraging them to set their sights on a career in science is the first step toward generating the space explorers of the future.

This project has been a labor of love for me and for our team of dedicated writers, researchers, artists, and designers. We have been delighted to witness the interaction between children and their parents, and the light of excitement in every family member's eyes as they experience amazing feats of past innovation and wondrous goals of the future. It is our sincere wish that as this exhibit moves from museum to museum, the interest in space that it inspires and its commitment to sharing the thrill of discovery permanently enhances each community.

We have produced this companion book so that you can take home with you the story of space exploration. We have asked renowned space historian Roger D. Launius, who served as chief historian for NASA before beginning his services at the National Air and Space Museum, to craft a story that is independent of the narrative in the exhibit. In this way, we aim to offer you a resource that builds on the information you encountered in the exhibit, allowing you and your family to further the process of discovery. To this end, Roger has also developed a helpful section of resources for additional information, which you will find in the back of this book.

We are honored that you have chosen to experience *Space: A Journey to Our Future*, and hope that you came away with a sense of the amazing possibilities the universe holds. Please continue on your journey of education and discovery in the following pages.

— *Stacy F. King, Clear Channel Exhibitions President and CEO*
January 2004, San Antonio, Texas

Saturn is not the only planet in our solar system to have rings around it, but our fifth planet's rings are by far the most impressive. This composite image shows Saturn as seen in many different wavelengths, when the planet's rings were at their maximum tilt of 27 degrees toward Earth — the best time for viewing Saturn's rings. Captured by the Hubble Space Telescope between March and April 2003, these images highlight more than one hundred bright and dark ringlets.

DARE TO DREAM
STUDYING THE HEAVENS

HUMAN CURIOSITY ABOUT THE UNIVERSE may be one of the few constants in recorded history. Ancient societies watched the heavens, building great observatories to chart the paths of the Sun, Moon, planets, and stars. The heavens became wrapped up in their religions, philosophy, and science. Deities from around the world have long been associated with celestial bodies, and virtually all pre-Christian religions equated objects in the night sky with supernatural forces. The Judeo-Christian tradition in Western civilization spent centuries trying to stamp out many of these beliefs — failing to do so entirely — and incorporated some celestial signs and symbols into its newer religious tradition. All human civilizations have pondered their place in the cosmos, sometimes viewing themselves as especially significant and placed at the universe's center by their god.

As scientific knowledge expanded, and the tools and methods for learning more about the heavens emerged, new understandings and perspectives replaced earlier ideas. All the contributors to this noble quest for wisdom have been innovators in a long tradition of scientific study that continues to the present. While we sometimes view earlier ideas about the cosmos as simplistic and ill-considered, they reflected what was then known about the subject. Centuries in the future, our descendents may laugh at what we today perceive as truths about the universe. Such, of course, is the nature of scientific inquiry.

The prehistoric people who built Stonehenge in England used observations of celestial bodies to chart planting seasons and measure other events, and assigned a religious significance to their studies. They created an astronomical calendar and predicted solar and lunar eclipses as well as the alignments of planets and constellations. To the ancient Egyptians, the Milky Way served as a "heavenly Nile" that helped explain the seemingly mysterious forces of nature. The Great Pyramid of Giza in Egypt, for example, is related to the cardinal points of the compass, and its corridors and shafts align with various stars and planets. The ancient Sumerians of 2500 B.C. kept precise observations of the movements of the Sun and Moon to create an accurate calendar, and astronomers in Babylon

Copernicus at Work by Matejko Jan (1873) depicts Nicolaus Copernicus (1473–1543), the founder of modern astronomy. Born in Poland, he studied the cosmos for many years before completing his great work, De Revolutionibus, in 1530. He asserted in it that the Earth rotated on its axis one time every day and completed a trip around the Sun every year, a correct but revolutionary concept for his time. De Revolutionibus launched a crisis in cosmology, as the Catholic Church insisted that the Earth rested at the center of the universe.

For thousands of years astron-omers have developed tools to make the discussion of the uni-verse easier to understand. One such tool, planispheres, were flat maps of the sky that could be rotated to help locate objects in the sky at any given time. Above is a planisphere of Ptolemy's geocentric universe as published in 1661. On the opposite page is a depiction of the Copernican cosmological system, also published in 1661, which replaced the Ptolemaic conception of the universe.

about 700 B.C. charted the paths of several planets and compiled observa-tions of fixed stars. Later they devised the zodiac, the first mechanism to divide the year into lunar periods.

Likewise, Chinese astronomers developed magnetic compasses and charted the stars for navigation. Pacific and Indian Ocean islanders, Persians, and Arabs also used knowledge gained about the stars and planets to navigate throughout their regions. In what became the Americas, ancient cultures also built astronomical observatories and observed in 1054, along with the Chinese, the explosion of a star

(called a supernova) that created the Crab Nebula, the hazy, bright cloud of gas that was shaped like a crab. The Mayans tracked the movements of the Moon, Venus, and other celestial bodies and used this information to develop complex calendars for planting and harvesting.

SUN AT THE CENTER

In ancient Greece, Aristarchus of Samos proposed a heliocentric (Sun-centered) universe. His contempo-raries did not accept his position, however, and his ideas went largely untaught for centuries. Even so, he

proved three well-known propositions about the universe: (1) the distance of the Sun from the Earth is greater than eighteen times, but less than twenty times, the distance of the Moon from the Earth; (2) the diameter of the Sun has that same proportion (eighteen to twenty times) to the diameter of the Moon; and (3) the ratio the diameter of the Sun has to the diameter of the Earth is greater than 19 to 3 but less than 43 to 6.

By A.D. 150 Aristarchus's ideas had been all but forgotten, replaced by those of Ptolemy of Alexandria, who had a geocentric (Earth-centered)

model of the universe. His great work, which he named the Almagest, placed the Earth at the center of the universe, with the Sun, stars, and planets embedded as jewels in a setting of crystalline spheres circling around it. As the Roman Cicero observed, "the motion of the spheres creates a harmony formed out of their unequal but well-proportioned intervals, combining various bass and treble notes into a melodious concert." The Ptolemaic system envisioned a geometric and fixed harmony of the universe and placed humanity squarely at its center.

This worldview did not significantly change until the sixteenth century, when Polish astronomer Nicolaus Copernicus saw that the flight of some planets could not be explained using the Ptolemaic idea of the universe. Accordingly, in the early 1500s he placed the Earth in orbit around the Sun. While Copernicus hesitated to overturn the Ptolemaic universe, others did not. Galileo Galilei used the newly invented telescope to show that the Earth and the planets revolved around the Sun. In 1616 Catholic clergy brought Galileo before the Inquisition in Rome, charging him

with heresy for challenging more than one thousand years of tradition about humanity as the center of the universe. By agreeing not to teach these ideas, he received no punishment, but in 1632 further charges compelled Galileo to recant his belief that the Earth revolved around the Sun. His legendary response, reported only later, was "*E pur si muove*" ("And it does move").

Regardless of heresy charges against Galileo, others took up his cause. Indeed, the "Copernican revolution" reshaped the human race's conception of the universe and our place in it. An Englishman, Isaac Newton, was by far

the most important of Galileo's inheritors, and by 1720 his three laws of motion had placed both astronomy and physics on a firm scientific foundation. He argued that universal gravitation, that is, the physical pull of all objects toward each other, accounted for the physical actions of celestial bodies. In particular, he demonstrated that the attraction of the Sun on a planet stood directly proportional to the product of the two masses and inversely proportional to the square of the distance separating them. Newton's ideas, so critical to the development of spaceflight, became the established

Ever since people realized in the 1600s that the Moon and planets are actual physical places, there existed a desire to go there. And it was artists and writers that first provided powerful visions of what those voyages of discovery might be like. Below left: This illustration from French science fiction writer Jules Verne's De la Terre à la Lune (From the Earth to the Moon) *(1865) depicts a giant cannon blasting a capsule to the Moon. Below center: This* *frame from French director Georges Méliès' best-known film,* Le Voyage dans la lune (A Trip to the Moon) *(1902), depicts a rocket hitting the Moon squarely in the eye. Right: Stanley Kubrick's* 2001: A Space Odyssey *(1968) set a new standard for science fiction film with its accurate depiction of space shuttles, a space station, a lunar base, and a human mission to Jupiter.*

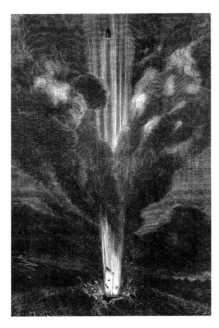

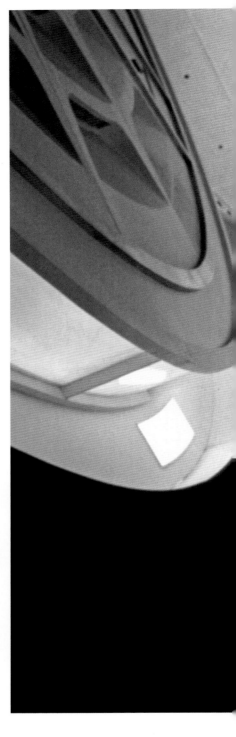

method of explaining the universe during his lifetime and remained so until the twentieth century.

BIRTH OF SCIENCE FICTION
Because of the "Copernican revolution," humanity for the first time began to think of the Moon and the planets as physical entities like the Earth, and with this reorientation that they might well harbor life that we could visit and that could visit us on Earth. This raised the possibility of space travel for the first time in human history. A flood of speculation about lunar flight followed. Johann Kepler, himself a ground-breaking astronomer, published a novel, *Somnium (Dream),* in

1634 that recounted a voyage to the Moon. He also used the latest scientific knowledge to speculate on the difficulties of overcoming the Earth's gravitational field, the nature of the elliptical paths of planets, the problems of maintaining life in the vacuum of space, and the geographical features of the Moon. This was the first, but far from the last, instance of scientists using fiction to advance their theories of the universe.

Other writings sparked by the "Copernican revolution" described fictional trips into space. Cyrano de Bergerac wrote *Histoire comique des etats et empire de la lune* (1649) — published in English as *The Comical History of the States and Empires of the Worlds of the Moon*

and Sun — describing several attempts to travel to the Moon. First, his hero tied a string of bottles filled with dew around himself, so that when the heat of the Sun evaporated the dew he would be drawn upward, but the hero only made it as far as Canada on that attempt. He finally reached the Moon by setting off firecrackers tied to his vehicle, becoming the first to use rocket thrust for space travel, anticipating Newton's third law about every action having an equal and opposite reaction.

Other writers adopted the conventions of fiction and fantasy to discuss the possibilities of space travel in the years that followed. Jules Verne and H. G. Wells are two prime examples.

Even as they sought to tell dramatic stories, they used their up-to-date scientific understandings to inform their speculations on the nature of other worlds. In 1865 Verne published *De la terre à la lune (From the Earth to the Moon),* containing scientific principles that were very accurate for the period. It described the problems of building a vehicle and launch mechanism to visit the Moon. Using a 900-foot-long cannon, Verne shot his characters into space. Verne picked up the story in a second novel, *Autour de la lune (Around the Moon),* describing a lunar flight. Wells published *War of the Worlds* in 1897 and *The First Men in the Moon* soon after. Both novels used sound sci-

entific principles to describe space travel and encounters with aliens.

More important, these and other science fiction writers offered a vision of spaceflight that inspired generations of scientists and engineers to make the space age a reality. Scattered throughout the world, they pioneered rocketry during the period before World War II. With the rise of fascism in the 1930s, however, many rocket scientists began working for their nation's militaries. By far the most significant example of this was the V-2 ballistic missile, developed by German engineer Wernher von Braun and first flown in October 1942. By the end of the war, 1,155 rockets had been fired against

England and another 1,675 had been launched against Antwerp and other continental targets. After World War II, the United States military undertook significant rocket development and fielded its first ballistic missiles in the 1950s. These rockets became the first used to launch astronauts into space. The connection between military operations and the development of rocket technology has been significant and longstanding.

SPACE RACE

The space age arrived on October 4, 1957, when the Soviet Union launched into orbit the first Earth satellite, *Sputnik 1*. Few present when *Sputnik*

burst onto the national scene failed to sense that something important had taken place. The Soviet Union had demonstrated significant technological ability, and it astounded the American public. In response, the United States undertook several critical efforts aimed at catching up with the Soviet Union's space achievements: a full-scale review of both the civil and military programs of the United States; the establishment of a presidential science advisor in the White House who had responsibility for overseeing the activities of the federal government in science and technology; the creation of the Advanced Research Projects Agency in the

Department of Defense; the establishment of the National Aeronautics and Space Administration (NASA) to manage civil space operations; and passage of the National Defense Education Act to provide federal funding for education in the scientific and technical disciplines.

Even though the space frontier opened as a result of a Cold War crisis, the seeds of acceptance had been a part of American popular culture for several years. Wernher von Braun burst onto the public stage in 1952 with a special issue of *Collier's* magazine on the possibilities of space flight. *Collier's* and von Braun came out with two more special issues during the next

A Vision Formed in a Cherry Tree

As a boy, Robert H. Goddard thrilled to the exploits described by Jules Verne and H. G. Wells. At the age of seventeen on October 19, 1899, he climbed a tall cherry tree behind the barn of the Goddard family house in Worcester, Massachusetts. He claimed to have a vision in the tree of "how wonderful it would be to make some device which had even the possibility of ascending to Mars." This event held a profound influence upon him for the rest of his life. "I was a different boy when I descended the tree from when I ascended," he recalled, "for existence at last seemed very purposive." Goddard would later

Robert H. Goddard at the blackboard in 1924

refer to October 19 as "Anniversary Day," and when in Worcester he never failed to revisit the tree.

The vision of space travel conjured in that cherry tree sustained Goddard through nearly a half century of activity that helped to open the space age. At his high-school oration in 1904 Goddard summarized his life's perspective: "It is difficult to say what is impossible, for the dream of yesterday is the hope of today and the reality of tomorrow." Goddard spent the rest of his life laying the groundwork for the space age. In 1919 the Smithsonian Institution published his classic study, *A Method of Reaching Extreme Altitudes*, which suggested that with a velocity of 6.95 miles per second an object could escape Earth's gravity.

On March 16, 1926, near Auburn, Massachusetts, Goddard launched his first liquid-fueled rocket. Rising only forty-three feet in 2.5 seconds, the rocket nonetheless heralded the modern age of rocketry. Cultivating benefactors to support his research, he then set up an experiment station near Roswell, New Mexico. Goddard did not live to see the dawn of the space age that he had helped fashion. He died in Annapolis, Maryland, on August 10, 1945.

two years. These framed the exploration of space in the context of the Cold War rivalry with the Soviet Union and concluded that "the time has come for Washington to give priority of attention to the matter of space superiority."

Following close on the heels of the *Collier's* series, Walt Disney Productions contacted von Braun to request his assistance in the production of three shows for Disney's weekly television series. The first of these, *Man in Space,* premiered on Disney's show on

March 9, 1955, with an estimated audience of forty-two million. The second show, *Man and the Moon,* also aired in 1955 and sported the powerful image of a wheel-like space station as a launching point for a mission to the Moon. The final show, *Mars and Beyond,* premiered on December 4, 1957, after the launching of *Sputnik 1.* Von Braun appeared in all three films to explain his concepts for human space flight, while Disney's characteristic animation illustrated the basic principles and ideas with wit and humor.

Because of such media efforts, during the decade following World War II, a sea change in American perceptions of the viability of space travel in the near-term took place. This can be seen in a December 1949 Gallup Poll in which only 15 percent of Americans believed humans would reach the Moon within fifty years, while 15 percent had no opinion and a whopping 70 percent believed that it would not happen within that time. In October 1957, at the same time as the launch of *Sputnik 1*, only 25 percent

believed that it would take longer than twenty-five years for humanity to reach the Moon, while 41 percent believed firmly that it would happen within twenty-five years and 34 percent were not sure. An important shift in perceptions had indeed taken place, and it resulted largely from well-known advances in rocket technology coupled with the efforts of such popularizers as von Braun in persuading the public of the feasibility of space flight.

Wernher von Braun became the public face of spaceflight for the United States during the 1950s. Here he is (front row, sixth from right) along with the rest of his "German rocket team" at White Sands Proving Grounds, New Mexico, in 1946. Far left: Von Braun poses with Walt Disney in 1954 during the filming of a television show on the space program. Left: An illustration shows a three-stage reusable launch vehicle that von Braun imagined could service an orbital space station.

The evolution of rocketry from Robert Goddard's first liquid-fueled double acting engine in 1926 to the mighty Saturn V Moon rocket in 1969 has been constant and impressive. Below left: Robert H. Goddard was the first to fly a liquid-fueled rocket, and here he poses with his double-acting engine, an important innovation that took more than two years to develop. It allowed him to launch the first successful liquid-propellant rocket on March 16, 1926. Below center: At Wallops Station on the eastern shore of Maryland, Durwood Dereng measures the elevation of a double Deacon booster rocket prior to the launch of an RM-10 research model built by the National Advisory Committee for Aeronautics (NACA) on February 6, 1951. Right: Also at Wallops Station, in 1959 NASA prepares to test "Little Joe," a test of the escape and recovery systems for the Mercury spacecraft that would launch Alan Shepard into space in 1961. Far right: This last image demonstrates the size of the Saturn V Moon rocket, which stood 363 feet high. Attached to the Launch Umbilical Tower, the Apollo 15 launch stack has just emerged from the Vertical Assembly Building (VAB), inching its way atop a massive crawler-transporter to Launch Complex 39.

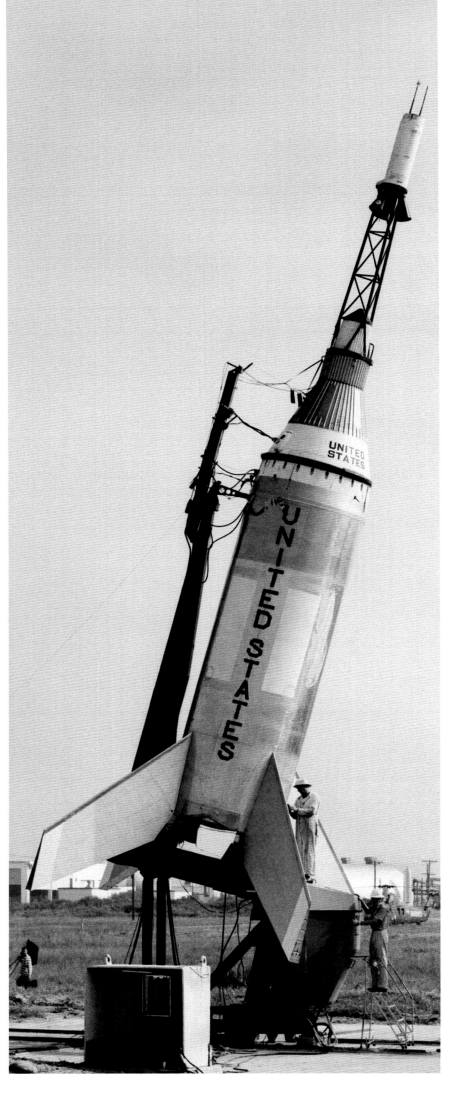

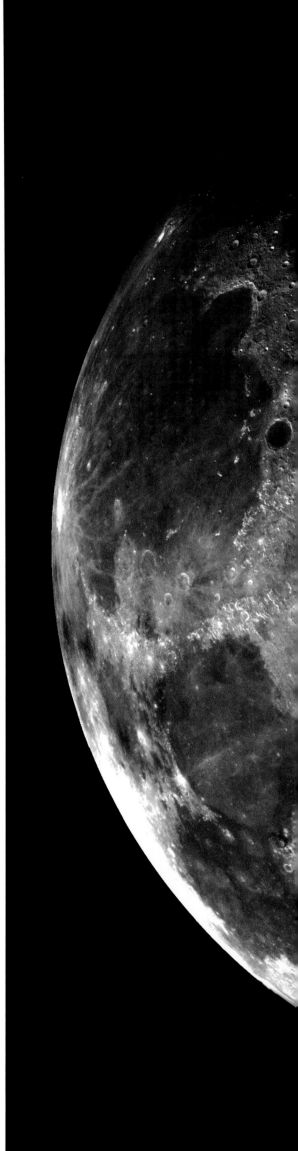

W HAT IS IT ABOUT THE MOON that captures the fancy of humankind? A silvery disk hanging in the night sky, it conjures up images of romance and magic. It has been counted upon to foreshadow important events, both of good and ill, and its phases served humanity for eons as its most accurate measure of time. Since ancient times, people have watched the Moon wax (appear to grow to a full circle) and wane (appear to shrink to but a sliver) and wondered at its beauty and mystery. The Moon holds an important place in many of today's religions, and it once played a part in other religions, such as Christianity, that no longer assign it special significance.

In the twentieth century, as spaceflight became a possibility, the Moon took on added meaning as Earth's nearest astronomical neighbor and a relatively easy destination for humankind to visit and explore. The Moon proved an early target for both the United States and the Soviet Union during the latter 1950s and the 1960s. The large number of spaceflight "firsts" associated with the Moon in the 1950s and 1960s clearly demonstrate the significant role lunar exploration played during that first heroic era of the space age. From the first clear photographs of the Moon to the Apollo landings to the robotic missions of the 1990s, the Moon maintains a hold on our imagination.

Americans sought the Moon first in 1958, eager to demonstrate leadership in space technology. The United States made four attempts to pilot the *Pioneer* spacecraft to the Moon in the winter of 1958–59, but each failed. The Soviets had more success. In January 1959 they sent *Luna 1* past the Moon and into orbit around the Sun, following up with *Luna 3* to transmit pictures of the far side of the Moon, giving the Soviets an important first in lunar exploration. Meanwhile, in March 1959, *Pioneer 5* finally flew past the Moon, much too late to assuage America's loss of pride and prestige. Thus ended the first phase of lunar exploration, with the Soviet Union the clear winner.

Later efforts had more success. In the 1960s NASA sponsored three major robotic programs to explore the Moon. The first, Project Ranger, saw its first six attempts fail. NASA then reorganized the Ranger project and did not try to launch

Humans have long equated the Moon with romance and mystery and desired to visit it. Here is a view of the Moon's north pole assembled from eighteen images taken by the Galileo *probe's imaging system as the spacecraft flew by on December 7, 1992. Only the left part of the Moon's surface shown here is visible from Earth.*

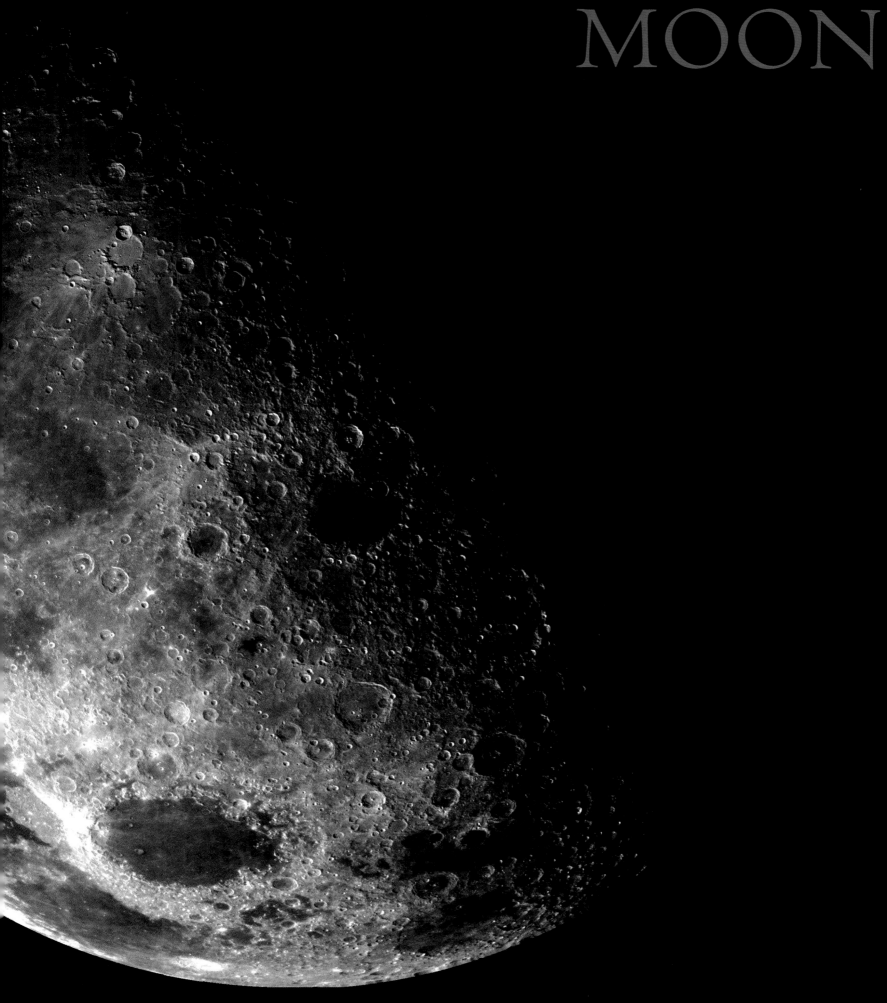

A DREAM COME TRUE

DESTINATION MOON

again until 1964. By that time its engineers had eliminated all the scientific instruments except a television camera. Going out in a blaze of glory, *Ranger* met its final objective by crashing into the Moon taking pictures. On July 31, 1964, the seventh *Ranger* worked and transmitted 4,316 beautiful high-resolution pictures of the Sea of Clouds. The eighth and ninth *Rangers* also worked well, thus ending the program on a high note.

Other lunar exploration projects followed, and they succeeded in providing data useful to both the scientific community and those planning human Moon landings as a part of the Apollo program. In 1965–66, five Lunar Orbiter probes took photos of the Moon's surface from orbit, imagery vital to planning for lunar landing missions. Likewise, in 1966–67, the Surveyor program soft-landed six probes on the Moon's surface (with one failure), providing

images and testing the characteristics of the lunar soil.

SETTING THE GOAL

Ranger, Lunar Orbiter, and Surveyor all aided in the largest and most stupendous space exploration ever undertaken, Project Apollo. The immediate reasons for the effort came in April 1961 when the United States suffered two foreign policy setbacks at the hands of the Soviet Union. The first of these was the Soviet Union's April 12 launch of Cosmonaut Yuri Gagarin on a one-orbit ride, and the second a disastrous U.S. invasion of the Bay of Pigs in Cuba, designed to overthrow Fidel Castro, less than a week later. On April 20, 1961, President John F. Kennedy dispatched a memorandum asking Vice President Lyndon B. Johnson to delineate a "space program which promises dramatic results in which we could win." LBJ met with political, foreign policy, scientific, and

technical advisors to craft a major policy decision in favor of a lunar landing program, with JFK announcing it to the world in a speech before a joint session of Congress on May 25, 1961. His words still capture the sense of awe and wonder of the moment: "I believe this nation should commit itself to achieving the goal, before this decade is out, of landing a man on the Moon and returning him safely to earth. No single space project in the period will be more impressive to mankind, or more important for the long-range exploration of space, and none will be so difficult or expensive to accomplish."

With the achievement of powered flight just over half a century old, Kennedy had made an audacious pronouncement. JFK announced his decision as a means of demonstrating the United States' technological virtuosity vis-à-vis the Soviet Union. In so doing Kennedy responded to perceived chal-

lenges to U.S. world leadership, not only in science and technology but also in political, economic, and especially military capability. Kennedy's decision immediately captured the American imagination, generating significant popular excitement that proved necessary for its long-term success. With this, Kennedy demonstrated his resolve to confront the Soviets in space. Thereafter, Apollo took on a life of its own and left an important legacy to both the nation and the proponents of space exploration.

For the next eleven years, Apollo consumed NASA's every effort. It required significant expenditures, costing $25.4 billion in 1960s dollars over the life of the program (more than $110 billion in 2003 dollars) to make it a reality. Only the building of the Panama Canal rivaled the Apollo program's size as the largest nonmilitary technological endeavor ever undertaken; only the Manhattan Project was comparable in a

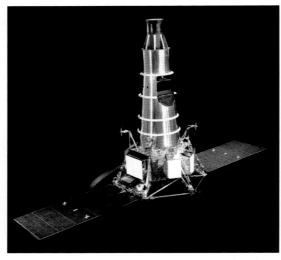

The Moon became the subject of intense interest as the space age opened, in part because of its close proximity to Earth but also because of its powerful hold on the human imagination. Right: The world's first view of Earth taken by a spacecraft from the vicinity of the Moon, coming from the United States's Lunar Orbiter 1 *on August 23, 1966. Above: Ham radio operators worldwide listened for signals from spacecraft seeking the Moon. Here a radio study group at a Moscow school tracks a satellite. Above center: The Ranger fleet of spacecraft launched in the mid-1960s provided for the first time over 17,000 images of the lunar* surface. They allowed scientists and engineers to study the Moon in greater detail than ever before, furthering knowledge necessary for the Apollo astronauts to successfully complete their landing missions. Above right: The Apollo 12 *spacecraft successfully landed within 600 feet of the robotic soft lander,* Surveyor 3, *in November 1969. This made it possible for astronauts Pete Conrad and Alan Bean to walk over to* Surveyor 3, *which had landed more than two years earlier on April 19, 1967, and reclaim its television camera and several other components for scientific analysis on Earth.*

wartime setting. Even NASA leaders expressed concern that it might prove too daunting a challenge. When Kennedy made his speech, Robert R. Gilruth, NASA's director of the Space Task Group charged with conducting human spaceflight, recalled that he was "aghast" at the lunar landing goal and what it would portend. After all, he reasoned, his organization now had to accomplish it. Rising to the challenge, project participants exhibited single-minded devotion to it for a decade.

The tight schedule for Apollo necessitated one of the three most significant decisions of the Apollo program, the lunar-orbit rendezvous mode decision in 1962. This involved launching the entire lunar spacecraft in one mission. It would then head to the Moon, enter into orbit, and dispatch a small lander to the lunar surface. It was the simplest of three possible methods in terms of both development and operational costs, but it was

risky. Since rendezvous would take place in lunar, instead of Earth, orbit, there was little room for error. Moreover, some of the trickiest course corrections and maneuvers had to be done after the spacecraft had been committed to a circumlunar flight.

Inside NASA, advocates of various approaches argued over the lunar mode while the all-important clock ticked. It was critical that NASA make a decision early, because the mode dictated the type of spacecraft built. Although NASA engineers could proceed with building the Saturn launch vehicle and define the basic components of the spacecraft — a habitable crew compartment, a service module containing propulsion and other expendable systems, and a Moon landing craft — they could not proceed much beyond rudimentary conceptions without a decision about the mode of transport. Using sophisticated technical analyses, NASA

wended its way toward a decision in 1962 in favor of lunar-orbit rendezvous. The last to give in was Wernher von Braun at the Marshall Space Flight Center during an all-day meeting on June 7. This set the stage for the development of hardware.

Even as these events took place, NASA worked to complete the program to put a man in space, named Project Mercury. Stubborn problems arose at almost every turn. The first spaceflight of an astronaut had been postponed for weeks so NASA engineers could resolve numerous details. It finally took place on May 5, 1961, less than three weeks before the Apollo announcement, and Alan Shepard undertook a suborbital flight of less than fifteen minutes. The second flight, in July 1961, another suborbital mission, had its hatch blow off prematurely, and *Liberty Bell* 7 sank into the Atlantic Ocean. Not until 1999 did a team of underwater archeolo-

gists bring up the spacecraft and restore it for public display. These suborbital flights, however, proved valuable for NASA technicians, who found ways to solve or work around literally thousands of obstacles.

The first American orbital mission took place on February 20, 1962, with John Glenn aboard, after several postponements. Making three orbits in *Friendship* 7, Glenn had his share of problems. He flew parts of the last two orbits manually because of an autopilot failure, and he left his retro-rocket pack (normally jettisoned) attached to the capsule during reentry because of a warning light indicating a loose heat shield. Despite this, Glenn's flight was successfully completed and provided a healthy boost in national pride. The public embraced Glenn as a personification of heroism and dignity. Hundreds of requests for personal appearances by Glenn poured into NASA headquar-

The man who gave Americans the Moon, President John F. Kennedy, left, in his historic message to a joint session of the Congress, on May 25, 1961, declared, ". . . I believe this nation should commit itself to achieving the goal, before this decade is out, of landing a man on the Moon and returning him safely to the Earth." This goal was achieved when astronaut Neil A. Armstrong became the first human to set foot upon the Moon at 10:56 P.M. EDT, July

20, 1969. Shown in the background is Vice President Lyndon B. Johnson. The man who made it possible to reach the Moon on JFK's time line was John C. Houbolt, above. Battling bureaucracy and conventional wisdom, Houbolt tirelessly argued in 1961 and 1962 that the best method of reaching the Moon was by a scenario called "lunar-orbit rendezvous."

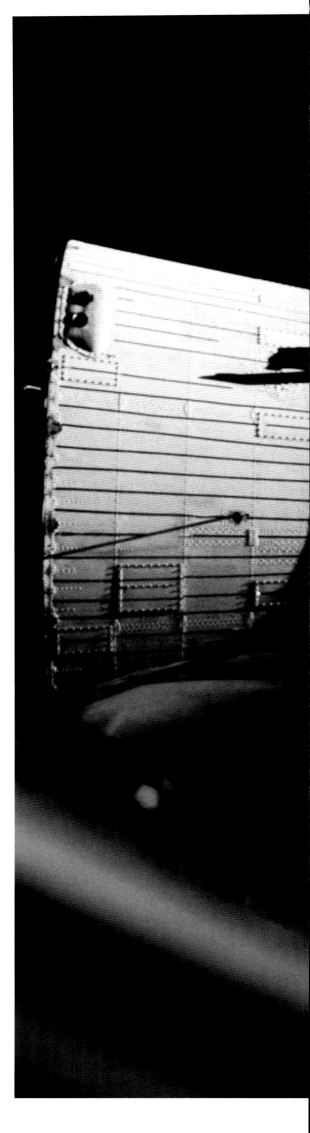

ters, and among other engagements the astronaut addressed a joint session of Congress and participated in several ticker-tape parades around the country. NASA thereby discovered a powerful public relations tool that it has employed ever since. Three more successful Mercury flights took place through May 15–16, 1963, when Gordon Cooper circled Earth twenty-two times in thirty-four hours.

PROJECT GEMINI

Even as this took place, NASA managers perceived a capability gap for human spaceflight between that acquired with Mercury and that necessary for a lunar landing. Thus Project Gemini, with two astronauts, was born. First, NASA needed to learn how to locate a spacecraft in orbit and to rendezvous and dock with it. Second, astronauts had to learn to leave the spacecraft and work in a vacuum, wearing newly developed spacesuits. Third, scientists had to determine if humans could survive in space for the two weeks required for

Apollo. Project Gemini filled this gap perfectly in 1965–66 with ten astronaut missions that set records for endurance, extravehicular activity (EVA) or spacewalks, and rendezvous and docking in orbit.

Meanwhile, NASA and its contractors struggled to build the most powerful and reliable rockets and spacecraft ever conceived. The mighty *Saturn V* Moon rocket stood 363 feet tall, with three stages, and generated 7.5 million pounds of thrust from five massive F-1 engines at launch. In an effort to prove this technology, NASA's Apollo program manager, Samuel C. Phillips, made the "gutsiest" call since the landing mode decision. Rather than test the program in stages, he adopted an "all up" concept to test the entire Apollo-Saturn system in flight. A calculated gamble, the *Saturn V* "all up" tests worked. In seventeen test launches and fifteen piloted launches, the Saturn booster family scored a 100-percent launch reliability rate.

Equally significant, the Apollo spacecraft proved a difficult engineer-

The Mercury Seven Astronauts, upper left, standing in front of a United States Air Force F-106B, were (from left to right) M. Scott Carpenter, L. Gordon Cooper, John H. Glenn Jr., Virgil I. "Gus" Grissom Jr., Walter M. "Wally" Schirra Jr., Alan B. Shepard Jr., and Donald K. "Deke" Slayton. Lower left: John Glenn during his first orbit in the Mercury Friendship 7 capsule on February 20, 1962. Lower right: Buzz Aldrin during an extravehicular activity while flying on Gemini XII in 1966. Aldrin, who trained in astrophysics at MIT, worked exhaustively during the Gemini program to develop the procedures necessary to maneuver in orbit. Right: In December 1965, Gemini VI and Gemini VII performed the first orbital rendezvous in space, a critical skill necessary to complete the Apollo program.

A Spokesman for Space

Wernher von Braun was one of the most important rocket developers and champions of space exploration during the 1930s through the 1970s. Born to a German noble in 1912, young von Braun became fascinated by the possibilities of space exploration by reading the science fiction of Jules Verne and H. G. Wells. As a teenager he joined in the German rocket society, *Verein für Raumschiffahrt,* and in 1932 went to work for the German army to develop ballistic missiles. He oversaw technical development of the V-2 ballistic missile. By the beginning of 1945 it was obvious to von Braun that Germany would not achieve victory against the Allies, and he arranged the surrender of five hundred of his best rocket scientists, along with plans and V-2 hardware, to the Americans. For fifteen years after World War II, he worked with the U.S. Army in the development of ballistic missiles.

Von Braun became one of the most prominent spokesmen of space exploration in the United States in the 1950s. In 1952 he gained renown as a participant in an important symposium dedicated to the subject, and he burst on the nation's stage

Von Braun (center) with President John F. Kennedy (pointing upward) and NASA Deputy Administrator Robert C. Seamans.

with a series of articles in *Collier's,* a popular weekly periodical of the era. He also became a household name following his appearance on three Disney television shows dedicated to space exploration in the mid-1950s. He was largely responsible for the first American orbital satellite, *Explorer 1,* which he launched on January 31, 1958, atop a ballistic missile he had developed. He also headed NASA's Marshall Space Flight Center in Huntsville, Alabama, throughout the 1960s, where he oversaw the building of the mighty *Saturn V* Moon rocket, the most powerful launcher ever developed. Long after his death in 1977, he remains one of the most recognizable names associated with space exploration.

ing challenge. On January 27, 1967, three astronauts scheduled to fly the first Apollo mission — Gus Grissom, Ed White, and Roger Chaffee — were aboard running a mock launch sequence. After several hours of work, a flash fire in the pure oxygen atmosphere intended for the flight engulfed the capsule and the astronauts asphyxiated. It took the ground crew five minutes to open the hatch because of internal pressures. NASA worked quickly to redesign the spacecraft, changing the 100-percent oxygen environment to a less flammable two-gas system. The final flight checkout of the spacecraft took place during *Apollo 7* on October 11–22, 1968, with astronauts Wally Schirra, Donn Eisele, and Walt Cunningham.

The next mission, *Apollo 8,* went to the Moon, representing a third critical decision for the Apollo program. At first planning to test Apollo hardware in the relatively safe confines of low Earth orbit, NASA engineers pressed for approval to make it a circumlunar flight as a public demonstration of what the United States could achieve. So far Apollo had been all promise; after the November 1968 decision to reconfigure the mission into a circumlunar flight, the delivery was about to begin.

Just before the holidays in 1968, the crew of *Apollo 8* — Frank Borman, Jim Lovell, and Bill Anders — focused a portable television camera on Earth. For the first time humanity saw its home from afar: a tiny, lovely, and fragile "blue marble" hanging in the blackness of space. When *Apollo 8* reached the Moon on Christmas Eve, this image of Earth was reinforced when the crew sent pictures of the planet back while reading from the

This view from the Apollo 11 *spacecraft shows Earth rising above the Moon's horizon. The lunar terrain pictured is in the area of Smyth's Sea on the near side. These images of Earth from the Moon gave humans a new perspective about our planet's place in the cosmos and have been important in the birth of the modern ecological movement as a symbol of what humanity must preserve.*

Book of Genesis and sending holiday greetings to the "people of the good Earth." Returning to Earth on December 27, *Apollo 8* represented a wonderful present to humanity after a devastating year in which American society registered crisis over Vietnam, race relations, urban problems, and a host of other difficulties. Two more Apollo missions, which tested the Lunar Module and the spacesuits needed to protect the astronauts, occurred before the climax of the program.

FIRST LUNAR LANDING

The first lunar landing came during the flight of *Apollo 11,* which lifted off on July 16, 1969, and reached the lunar surface at 4:17 P.M. EST on July 20, 1969. The Lunar Module — with astronauts Neil Armstrong and Buzz Aldrin aboard — landed on the lunar surface, while Michael Collins orbited overhead in the Apollo Command Module. After checkout, Armstrong set foot on the surface, telling millions who saw and heard him on Earth that it was "one small step for [a] man — one giant leap for mankind." Aldrin soon followed him out, aptly describing the Moon as "magnificent desolation." The two bunny-hopped around the landing site in the one-sixth lunar gravity, planted an American flag (but omitted claiming the land for the United States as European countries had routinely done during exploration), collected soil and rock samples, and set up scientific experiments. The next day they launched back to the Apollo capsule orbiting overhead and began the return trip to Earth, splashing down in the Pacific Ocean on July 24.

They had their share of difficulties accomplishing all this. The autopilot

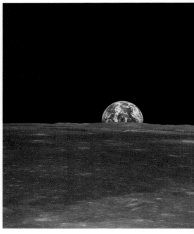

The awesome power of the mighty Saturn V lifts the Apollo 11 spacecraft and its crew of Neil Armstrong, Buzz Aldrin, and Michael Collins into Earth orbit at 9:32 A.M. EDT, on July 16, 1969, from the Kennedy Space Center's Launch Complex 39A in Florida, left. Below left: Members of the Kennedy Space Center launch control team rise from their consoles to watch the Apollo 11 liftoff through a window. The Apollo 11 first stage generated more than 7.5 million pounds of thrust as the thirty-six-story space vehicle began its journey to the Moon. Below right: The liftoff was caught on a 70mm telescope camera mounted in a pod on a cargo door of an Air Force EC-135N aircraft, flying at an altitude of 35,000 feet. Overleaf: While on the surface of the Moon on July 20, 1969, Neil Armstrong — seen in shadow in the left foreground — took this partial pan on the rim of East Crater, so named because it was sixty meters east of the Lunar Module. He had flown over this thirty-meter crater during the final approach and, near the end of the time scheduled to be on the lunar surface, ran out to it to take these pictures.

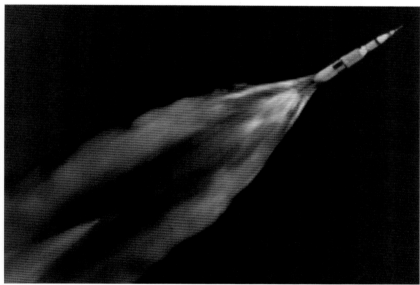

of the Lunar Module, a complex but delicate vehicle that could not withstand Earth's gravity and atmosphere, attempted to land in a boulder-strewn crater, so Armstrong took control and set the craft down safely with less than thirty seconds of fuel remaining. When Armstrong sent this calm message, "Houston. Tranquility Base here. The Eagle has landed," the world breathed a collective sigh of relief. The response from Mission Control betrayed the excitement and concern on the ground: "Roger, Tranquility. We copy you on the ground. You've got a bunch of guys about to turn blue. We're breathing again. Thanks a lot."

Armstrong and Aldrin also used the lunar EVA suits for the first time, a one-piece suit custom-tailored for each astronaut over several layers of undergarments, all made from materials like kapton, mylar, cotton, and nylon. The suit interfaced with the Portable Life Support System (PLSS) and could maintain comfort for astronauts in temperatures of -290 to +310 degrees Fahrenheit. Modified over several years, the suit proved remarkably efficient in protecting astronauts and making it possible for them to conduct lunar exploration.

A Rousing Success

Apollo 11 elicited an ecstatic reaction around the globe as everyone shared in its success. Ticker-tape parades, speaking engagements, public relations events, and a world tour by the astronauts helped create good will both in the United States and abroad. President Richard Nixon, for example, told an assembled audience that the flight of Apollo 11 represented the most significant week in the history of Earth since the creation. Clearly, the president viewed the endeavor as groundbreaking and permanent, a legacy of accomplishment that future generations would reflect on as they plied space and colonized planets throughout the galaxy. Hyperbole? Of course, but Nixon's comment reflected the exuberance of the moment.

Six more Apollo missions followed at approximately six-month intervals through December 1972, and with the exception of the aborted Apollo 13 flight, each of them increased the time spent on the Moon. Three of the later Apollo missions used a lunar rover vehicle to travel in the vicinity of the landing site, but none of them equaled the excitement of Apollo 11. The scientific experiments placed on the Moon and the lunar soil samples returned through Project Apollo have provided grist for scientists' investigations of the solar system ever since. Indeed, the scientific return was astounding.

There are three critical legacies of Project Apollo. First, it successfully accomplished the political goals that brought it about. Kennedy undertook Apollo to deal with a Cold War crisis in 1961. At the time of the Apollo 11 landing, Mission Control in Houston flashed the words of JFK announcing the Apollo commitment on its big screen. Those phrases were followed with "TASK ACCOMPLISHED, July 1969." While the scientific return of the Apollo program proved important, America did not go to the Moon in the 1960s and 1970s out of scientific curiosity. Any assessment that does not recognize the political success of landing on the Moon by the end of the 1960s is incomplete, for that was Apollo's primary goal.

Second, Apollo was a triumph of large-scale technical management. James E. Webb, the NASA administrator at the height of the program

The Loneliest Job in the Solar System

Michael Collins became an astronaut in the third group selected by NASA in 1963. Known as quiet and reflective, Collins piloted *Gemini 10* in 1966, and remained in the *Apollo 11* Command Module in orbit around the Moon when his colleagues, Neil Armstrong and Buzz Aldrin, landed on the lunar surface in 1969. During the *Apollo 11* mission, Collins had the loneliest job in the solar system as he orbited the Moon by himself.

In the most honest and reflective of all the astronaut memoirs, *Carrying the Fire* (1974), Collins writes of his solitude in lunar orbit. As he disappeared on the backside of the Moon from Earth, he recalled, "I am alone now, truly alone, and absolutely isolated from any known life, I am it. If a count were taken, the score would be three billion plus two over on the other side of the moon, and one plus God only knows what on this side. I feel this powerfully—not as fear or loneliness—but as awareness, anticipation, satisfaction, confidence, almost exultation. I like the feeling. Outside my window I can see stars—and that is all. Where I know the Moon to be, there is simply a black void, the Moon's presence is defined solely by the absence of stars." He compared it to being in a skiff in the middle of the ocean with only the stars above and black water below. It proved a profoundly moving experience for him.

Michael Collins left NASA in 1970 and became the first director of the Smithsonian Institution's National Air and Space Museum, continuing to write eloquently of the possibilities of spaceflight. Among other works he published were *Liftoff: The Story of America's Adventure in Space* (1988) and *Mission to Mars* (1990), a powerful exposition on the value of a human mission to Mars.

Neil Armstrong, Michael Collins, and Buzz Aldrin prior to their Apollo 11 *mission*

between 1961 and 1968, always contended that Apollo represented a management challenge more than anything. Establishing procedures to deal with thousands of personnel and components may have been the most significant result of the program.

Third, Project Apollo forced the people of the world to view the planet Earth in a new way. For example, Anne Morrow Lindbergh suggested in *Earthshine* (1969) that humanity gained a "new sense of awe and mystery in the face of the vast marvels of the solar system." She added, "Man had to free himself from earth to perceive both its diminutive place in the solar system and its inestimable value as a life-fostering planet." The writer Archibald MacLeish said it this way: "To see the Earth as it truly is, small and blue and beautiful in that eternal silence where it floats, is to see ourselves as riders on the Earth together, brothers on that bright loveliness in the eternal cold — brothers

who know now that they are truly brothers." This change in perceptions helped to galvanize the modern environmental movement.

Apollo proved so successful that the United States did not think it necessary to return to the Moon until 1994, when the Clementine orbiter mapped the lunar surface and returned data that suggested that ice might exist under the Moon's polar cap. Excitement over this discovery spurred those developing Lunar Prospector to make the search for ice a central part of its mission. Launched on January 6, 1998, Lunar Prospector globally mapped the Moon, confirming the Clementine polar data. From ice, humans could create water, oxygen, and hydrogen. This finding makes human colonization of the Moon possible, thereby exciting many about a future lunar colonization effort.

One of the first footprints left on the Moon, Buzz Aldrin's bootprint, above, was made during the Apollo 11 *mission on July 20, 1969, above. Aldrin took several of these images as part of an experiment to test the characteristics of lunar soil, but the images quickly became icons of the human quest to explore. Right: Neil Armstrong took this photograph of Buzz Aldrin on July 20, 1969. Armstrong is reflected in Aldrin's visor, as is a leg of the "Eagle" lunar module and the few traces of humankind at Tranquility base. Overleaf: Scientist-astronaut Harrison H. Schmitt is photographed standing next to a split boulder during the third* Apollo 17 *extravehicular activity (EVA-3) at the Taurus-Littrow landing site on the Moon in December 1972.*

EXPLORING THE SOLAR SYSTEM

This stunning panorama of the Mars Pathfinder landing site of July 1997 was created by combining a mosaic of images from the mission with photographs of a full-scale museum model of the Mars Pathfinder lander. The image shows the major features of "Rock Garden," the rover Sojourner (at top), and the rock Yogi (next to Sojourner).

S INCE THE 1950s humans have sent robot explorers to every planet of the solar system, with the exception of Pluto. We have placed spacecraft in orbit around our Moon and the planets Venus, Mars, and Jupiter and have landed them on our Moon, Mars, and Venus. Stunning missions to explore the outer solar system have yielded a treasure of knowledge about our universe, how it originated, and how it works. Missions to Mars have shown the powerful prospect of past life on the red planet. Missions to Venus, including some that landed on it, and to Mercury have harvested new understandings about the inner planets. Most important, we have learned that, while all the other planets in our system seem hostile, Earth is a place where everything necessary to sustain our form of life is just right.

Of course, planetary exploration did not take place by magic. Like other aspects of the space program, it began as a race between the United States and the Soviet Union. Afterward, it expanded in scope as scientists created two types of spacecraft: the first, a probe sent toward a heavenly body, and the second, an Earth-orbiting observatory. The knowledge gained from this new data revolutionized humanity's understanding of Earth's planetary neighbors. These studies and photographs of the planets and theories about the origins of the solar system, perhaps as much even as Project Apollo, captured the imagination of people from all backgrounds and perspectives. As a result, NASA had little difficulty in capturing and holding a widespread interest in this aspect of the space science program.

A centerpiece of NASA's planetary exploration effort in the 1960s was the Mariner program, which originated in the early part of the decade to investigate Mars and Venus. Mariner satellites proved enormously productive throughout the 1960s. Mariner made a huge impact in the early 1960s as the first to reach Venus. Both the evening and the morning star, Venus had long enchanted humans as a planet shrouded in a mysterious cloak of clouds. It was also the closest planet to Earth, and a near twin in terms of size, mass, and gravitation.

After Earth-based efforts to study the planet using radar in 1961 (from which we learned, among other things, that Venus rotated in the opposite direction of its orbital motion), both the Soviet Union and the United States began a robotic

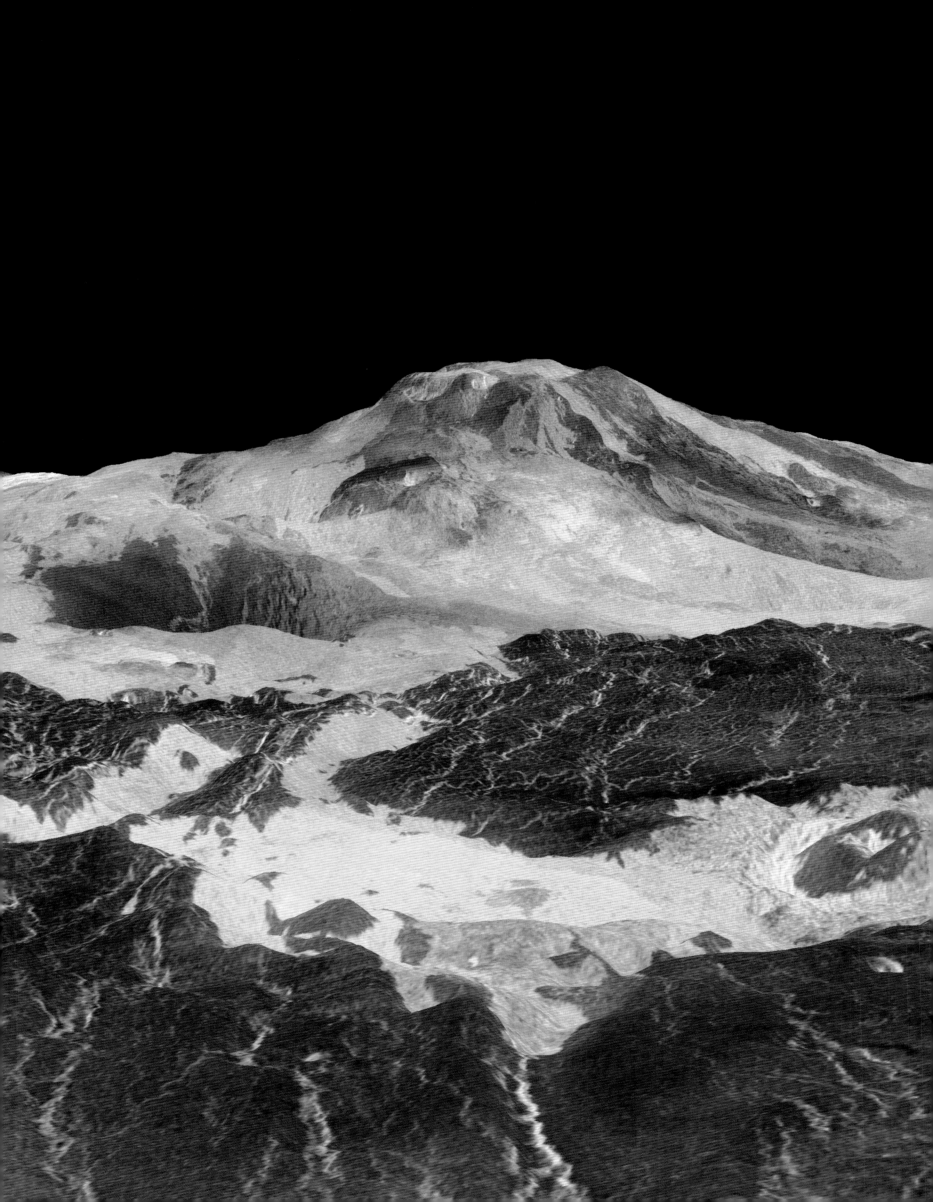

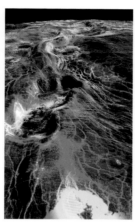 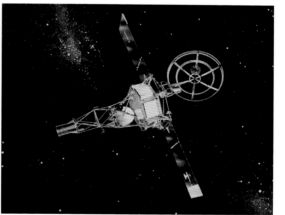 

Beginning in 1989 the Magellan *probe to Venus revealed the planet as never before using radar topographic imagery. At left, the volcano Maat Mons rises nearly three miles above the surrounding Venusian terrain. Above: This computer-generated image of Venus based on* Magellan *data shows part of the lowland plains in Sedna Planitia with blue representing the lowest areas and red the highest. Above center:* Mariner 2 *was the world's first successful interplanetary spacecraft. Launched August 27, 1962, on an Atlas-Agena rocket, it passed within about 21,000 miles of Venus, sending back new information about interplanetary space and the Venusian atmosphere. Above right: The* Magellan *spacecraft with its attached Inertial Upper Stage booster is in the Space Shuttle* Atlantis *payload bay prior to launch in 1989.*

race there. The United States claimed the first success in planetary exploration during the summer of 1962 when *Mariners 1* and *2* were launched toward Venus. After losing *Mariner 1* during launch, NASA sent out *Mariner 2*, which flew by Venus on December 14, 1962, at a distance of 21,641 miles. It probed the clouds, estimated planetary temperatures, measured the charged particle environment, and looked for a magnetic field similar to Earth's Magnetosphere (but found none). After that encounter, *Mariner 2* sped inside the orbit of Venus and eventually ceased operations on January 3, 1963, when it overheated. In 1967 the United States sent *Mariner 5* to Venus to investigate the atmosphere. Both spacecraft demonstrated that Venus was a very inhospitable place for life to exist, determining that the entire planet's surface was a fairly uniform 800 degrees Fahrenheit, thus ending the probability that life — at least as humans understood it — existed there.

The most significant mission to Venus took place when the *Magellan*

orbiter mapped the planet with imaging radar. This mission followed a *Pioneer Venus 1* spacecraft that had orbited the planet throughout the 1980s, completing a low-resolution radar topographic map, and *Pioneer Venus 2*, which had dispatched heat-resisting probes to penetrate Venus's dense clouds. It also built on knowledge acquired by the Soviets, who had compiled radar images of the northern part of Venus and had deployed balloons into the Venusian atmosphere. *Magellan* arrived at Venus in September 1990 and mapped 99 percent of the surface at high resolution, parts of it in stereo. This data betrayed some surprises: among them the discovery that plate tectonics was at work on Venus and that lava flows showed clearly the evidence of volcanic activity. In 1993, at the end of its mission, scientists turned their attention to a detailed analysis of *Magellan*'s data.

THE GRAND TOUR

During the 1970s, NASA's leadership hatched one of its most audacious space science missions. Once every 176

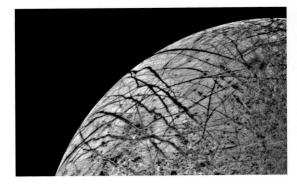

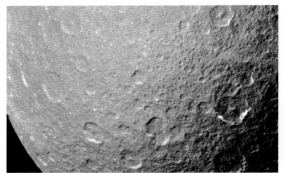

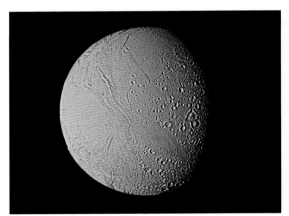

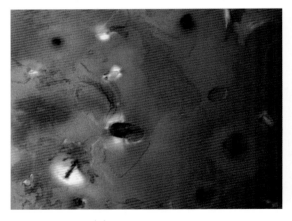

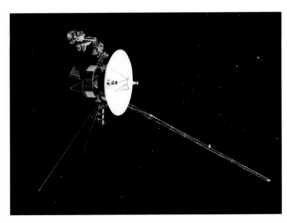

years, the giant planets of the outer solar system gather on one side of the Sun, and such a configuration was due. This geometric line-up made possible close-up observation of all the planets in the outer solar system (with the exception of Pluto) in a single flight, the so-called "Grand Tour." A flyby of each planet would bend the spacecraft's flight path and increase its velocity enough to deliver it to the next destination. This would occur through a complicated process known as "gravity assist," something like a slingshot effect, whereby the flight time to Neptune could be reduced from thirty to twelve years.

To prepare the way for the Grand Tour, NASA conceived *Pioneers 10* and *11* in 1964 as outer solar system probes. Launched in 1973, these probes toured Jupiter and Saturn and continued beyond the solar system. Both *Pioneers* were remarkable space probes, stretching a thirty-month-design life cycle to a mission of more than twenty years and returning useful data not just about the outer planets, but also about some of the mysteries of interstellar space.

Meanwhile, NASA technicians prepared to launch what became known as Voyager. In 1977 NASA launched *Voyagers 1* and *2*, which are still returning astonishing data. They first flew by Jupiter and Saturn, just as the earlier *Pioneers 10* and *11* had done, and then *Voyager 1* went on to Uranus and Neptune. Eventually, *Voyagers 1* and *2* explored all the giant outer planets, forty-eight of their moons, and the unique systems of rings and magnetic fields those planets possess. The two spacecraft returned information that revolutionized solar-system science. The two Voyagers took well over 100 thousand images of the outer planets, rings, and satellites, as well as millions of magnetic, chemical spectra, and radiation measurements. They discovered rings around Jupiter, volcanoes on Io, shepherding satellites in Saturn's rings, new moons around Uranus and Neptune, and geysers on Triton.

LOCATING WATER

It was nearly two decades after Voyager before any spacecraft ventured to the outer solar system again. In October

Voyagers 1 and 2 captured an array of stunning images during their travels. Far left: Propelled into space atop a Titan / Centaur rocket, Voyager 2 was launched on August 20, 1977, from the NASA Kennedy Space Center at Cape Canaveral in Florida. Near left: An artist's concept of Voyager 1 as it traveled on its "Grand Tour" of the outer solar system. At right, top to bottom: Voyager 2 captured cracks in the ice that covered water on Europa, one of Jupiter's satellites. A false color picture of another of Jupiter's moons, Callisto, alongside a Voyager 2 color photo of Ganymede, the largest satellite discovered by Galileo. This view of Enceladus, the orange moon of Saturn, was created from several images obtained on August 25, 1981. Voyager 1 observed several active volcanoes and photographed this South Polar region of Jupiter's moon Io in 1980. Voyager 1 took this high-resolution color image of Rhea just before the spacecraft's closest approach to the Saturnine moon on November 12, 1980. A mosaic of the four highest-resolution images of Ariel, a satellite of Uranus, is alongside a photograph of Uranus's outermost moon, Oberon. Opposite page, top left: Voyager 1 captured this exciting image of Saturn and two of its moons, Tethys and Dione, on November 3, 1980, from eight million miles. Top right: Voyager 2's camera took this image on July 3, 1979, showing the white oval situated south of the Great Red Spot. Bottom: This picture of Neptune came from Voyager 2 of Uranus.

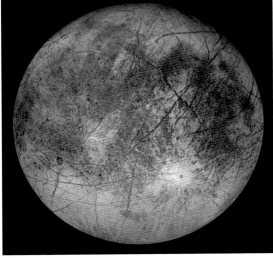

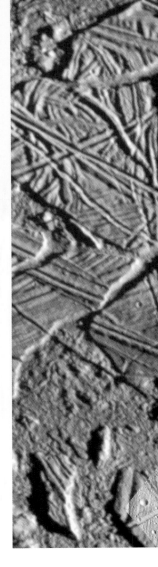

This photograph, below, of asteroid 951 Gaspra is a mosaic of two images taken by the Galileo spacecraft from a range of 3,300 miles on October 29, 1991. Gaspra's highly irregular shape suggests that the asteroid was derived from a larger body by a series of catastrophic collisions. Right: Three views of Europa, revealing that its surface strongly resembles images of sea ice on Earth and may indicate the possibility that aquatic life might exist there. This is reinforced by the far-right image, which shows irregularly shaped blocks of water ice shifted, rotated, and even tipped and partially submerged within a mobile material that was either liquid water, warm mobile ice, or an ice-and-water slush.

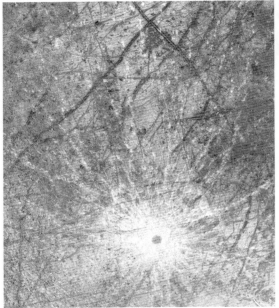

1989, NASA's *Galileo* spacecraft began a gravity-assisted (using the gravity of other planets to "slingshot" it toward its destination, saving fuel) journey to Jupiter, where it sent a probe into the atmosphere.

In mid-1995 *Galileo* deployed the probe that would parachute into Jupiter's dense atmosphere. The two spacecraft flew in formation the rest of the way to Jupiter, and while the probe began its descent into the atmosphere, the main spacecraft went into a trajectory that placed it in a near-circular orbit. On December 7, 1995, the probe began its descent. Its instruments began relaying data back to the orbiter on the chemical composition, the nature of the cloud particles and structure of the cloud layers, the atmosphere's radiative heat balance and pressure and dynamics, and the ionosphere. The probe lasted for about forty-five minutes before the atmosphere and the pressure of the planet destroyed

it. *Galileo*'s measurements have brought a reinterpretation of human understanding about Jupiter and its moons.

Most significant in terms of results has been the discovery of a frozen ocean of water covering Europa, one of the principal moons of Jupiter. On August 13, 1996, data from *Galileo* revealed that Europa may harbor "warm ice" or even liquid water — a key element in life-sustaining environments. Many scientists and science fiction writers have speculated that Europa — in addition to Mars and Saturn's moon Titan — is one of the three planetary bodies in this solar system that might possess, or may have possessed, an environment where life could exist. *Galileo*'s photographs of Europa revealed what appeared to be ice floes similar to those seen in Earth's polar regions. The pictures also revealed giant cracks in Europa's ice where warm-water "environmental niches" may exist. Early the next year *Galileo* discovered icebergs on Europa, lending credence to the possibility of hidden,

subsurface oceans. These findings generated new questions about the possibility of life on Europa. With the mission's end in 2003 NASA plunged *Galileo* into Jupiter, the probe transmitting its last data as it headed into the mighty planet's clouds.

In 2000–2001, NASA's Near Earth Asteroid Rendezvous (NEAR) mission to the asteroid Eros also achieved excellent results. NEAR was the first spacecraft to orbit an asteroid. It met all of its scientific goals in its year of orbiting the asteroid Eros and then undertook a landing on the surface on February 12, 2001. The chief goal of the landing was to gather close-up pictures of the boulder-strewn surface of 433 Eros, more than 196 million miles from Earth. During its five-year, two-billion-mile journey, the NEAR mission provided the most detailed profile yet of a small celestial body.

Finally, in 1997 NASA launched the *Cassini* spacecraft on its voyage to Saturn. In some respects a sister space-

craft to the remarkable *Galileo* vehicle at Jupiter, this spacecraft will linger in the Saturn system for several years collecting all manner of data about Saturn and its moons. Once it arrives, *Cassini* is expected to provide a similar level of stunning scientific data about Saturn's system to the rich harvest that *Galileo* brought to the human race about Jupiter and its moons.

UNLOCKING MARS'S MYSTERIES
From the beginning of the space age Mars has attracted significant attention, an attention it has yet to relinquish, prompting many missions there as well. In July 1965 *Mariner 4* flew by Mars, taking twenty-one close-up pictures. *Mariners 6* and *7* verified the Moon-like appearance of Mars and gave no hint that it had ever been able to support life. Among other discoveries from these probes, it was found that much of Mars was cratered like the Moon, that volcanoes had once been active on the planet, that the frost

Carl Sagan holding pictographs affixed to
Pioneers 10 *and* 11

Billions and Billions of Stars

Carl Sagan played a leading role in spaceflight from its earliest years, and no one has been more important as a public voice of space exploration since the 1970s. His powerful intellect, striking charm, and enormous charisma enabled him to become the public's intellectual in matters of space science and exploration. But he also engaged in strictly scientific pursuits. He worked with NASA beginning in the 1950s, taught Apollo astronauts about lunar science, and labored as a scientist on the Mariner, Viking, Voyager, and Galileo expeditions to the planets. He helped solve the mysteries of the high temperature of Venus (a massive greenhouse effect), the seasonal changes on Mars (windblown dust), and the reddish haze of Titan (complex organic molecules). Fittingly, Asteroid 2709 was named for him.

As a Pulitzer Prize–winning author, Sagan wrote many bestsellers, including *Cosmos*, which became the best-selling science book ever published in English. This work accompanied his Emmy and Peabody Award–winning television series, seen by 500 million people in sixty countries. At the time of his death on December 20, 1996, Sagan served as the David Duncan Professor of Astronomy and Space Sciences and Director of the Laboratory for Planetary Studies at Cornell University. His last book, *The Demon-Haunted World: Science as a Candle in the Dark*, was released by Random House in March 1996. Appropriately, this book offered science-based analysis of commonly held supernatural occurrences. Sagan also co-produced and co-wrote the acclaimed film *Contact*, based on his powerful novel of humanity's encounter with extraterrestrial life, a search for which he dedicated his life.

observed seasonally on the poles was made of carbon dioxide, and that huge plates indicated considerable tectonic activity. *Mariner 9* proved more interesting in November 1971 when it detected Nix Olympia (Snows of Olympus). It also found the remains of giant extinct volcanoes dwarfing anything on Earth. Mons Olympus was three hundred miles across at its base, with a crater in the top forty-five miles wide. Rising twenty miles from the surrounding plane, Mons Olympus was three times the height of Mount Everest. Later pictures also showed a canyon, Valles Marineris, 2,500 miles long and 3.5 miles deep. Later meandering "rivers" appeared, indicating that at some time in the past water had probably flowed on Mars. The discovery excited a flurry of interest in the planet.

Viking, a very important mission to Mars, used two identical spacecraft consisting of a lander and an orbiter. Launched in 1975 from the Kennedy Space Center, Florida, *Viking 1* spent

Mars was observed by Mariner 9 *on November 11, 1971, at a range of about 408,000 miles, right. Below: Olympus Mons (Nix Olympica, or Snows of Olympus) as photographed by* Mariner 9 *on November 27, 1971. The tallest volcano in the solar system, three times higher than Earth's Mount Everest, Olympus Mons can exist only because of the low gravity of Mars and the lack of a surface tectonic motion.*

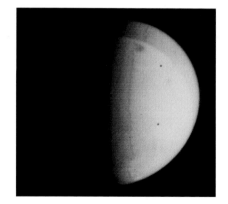

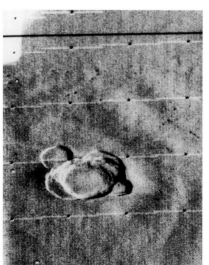

Our first representative on Mars, Sojourner

The Little Probe That Could

The Sojourner rover, a lightweight robot on wheels, accomplished in 1997 a revolutionary feat on the surface of Mars. For the first time, a thinking robot equipped with sophisticated laser eyes and automated programming reacted to unplanned events on the surface of another planet. Sent to the red planet aboard the *Mars Pathfinder, Sojourner* was ordered by NASA controllers to began making some of its own decisions within a few days of reaching the Martian surface. In so doing, *Sojourner* made trips between designated points without the benefit of detailed information to warn it of obstacles along the way. Moving slowly at 1.5 feet per minute, *Sojourner* stopped along the way to sense the terrain and process information; in so doing it added significantly to the scientific return of the Pathfinder mission.

Sojourner was designed by a large NASA team led by Jacob Matijevic and Donna Shirley. In all, it traveled about 328 feet in 230 commanded maneuvers, performed more than sixteen chemical analyses of rocks and soil, carried out soil mechanics and technology experiments, and explored about 2,691 square feet of the Martian surface. Returning an unprecedented 2.3 gigabits of data and running twelve times longer than its expected lifetime of seven days, *Sojourner* lost communication with the flight team on September 27, 1997.

In a contest, Valerie Ambroise of Bridgeport, Connecticut, then age twelve, submitted the winning essay recommending the robot be named after Sojourner Truth, an African-American abolitionist and champion of women's rights. Sojourner Truth made it her mission to "travel up and down the land," advocating the ending of slavery and the rights of women.

nearly a year cruising to Mars, placed an orbiter in operation around the planet, and landed on July 20, 1976, on the Chryse Planitia (Golden Plains). Following an identical trajectory, *Viking 2* landed on September 3, 1976. One of the most important scientific activities of Viking probes involved an attempt to determine whether life had existed on Mars. Mission biologists concluded that the combination of solar ultraviolet radiation that saturates the surface, the extreme dryness of the soil, and the oxidizing nature of the soil chemistry had prevented the formation of living organisms in the Martian soil. However, the question of life on Mars at some time in the distant past remains an unanswered question.

Since the Viking landings, there have been several missions to Mars seeking to unlock its mysteries. In 1996 a team of scientists announced that a Mars meteorite found in Antarctica contained possible evidence of ancient Martian life. They held that when the 4.2-pound, potato-sized rock identified as ALH84001 formed as an igneous rock about 4.5 billion years ago, Mars was much warmer and probably contained oceans hospitable to life. Then, about fifteen million

years ago, a large asteroid hit the red planet and jettisoned the rock into space, where it remained until it crashed into Antarctica about 11,000 B.C.E. Scientists presented evidence to suggest that ALH84001 contained fossil-like remains of Martian microorganisms dating back 3.6 billion years. The findings electrified the scientific world, which is still debating the evidence, and added support for an aggressive set of missions to Mars by the year 2010 to help discover the truth of these theories.

THE TRUTH ABOUT SOJOURNER

Feeding on the interest of the ALH84001 meteorite but not directly related to it, *Mars Pathfinder* landed on Mars on July 4, 1997. Its 23-pound robotic rover, named *Sojourner,* departed the main lander and began to record weather patterns, atmospheric opacity, and the chemical composition of rocks washed down into the Ares Vallis flood plain, an ancient outflow channel in Mars's northern hemisphere. Atmospheric-surface interactions, measured by instruments onboard the lander, confirmed some climatic conditions observed by *Viking* in the later 1970s, while raising questions about

other aspects of the planet's global system of transporting volatiles such as water vapor, clouds, and dust.

Another mission reached Mars on September 11, 1997, when *Mars Global Surveyor* entered orbit. Despite significant setbacks to the Mars exploration program with the failure of two missions in 1999, scientific returns from *Mars Global Surveyor* sustained interest in the planet. In what may prove a landmark discovery, scientists announced on June 22, 2000, that a feature observed on the planet suggested that liquid water almost certainly abundantly flowed in the planet's history. Scientists compared that feature to those left by flash floods on Earth.

Everyone agreed that the presence of liquid water on Mars had profound implications for the question of life on the red planet. NASA's Associate Administrator for Space Science, Ed Weiler, commented, "If life ever did develop there, and if it survives to the present time, then these landforms would be great places to look." The gullies observed in the images were on cliffs — usually in crater or valley walls — and showed a deep channel with a collapsed region at its upper end, and at the other end an area of accumu-

This image, left, is from Viking 2 on Utopia Planitia taken during the first Mars landing on November 23, 1976. Near right: The Sojourner rover that explored Mars in the summer of 1997. Far right: This is the first contiguous, uniform 360-degree color panorama taken by the Imager for Mars Pathfinder (IMP) in July 1997. In the right center is Sojourner, undertaking X-ray spectrometer analysis of the large rock, Yogi.

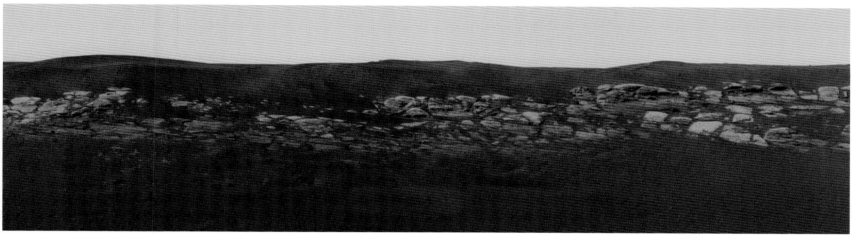

lated debris that appeared to have been transported down the slope. Relative to the rest of the Martian surface, the gullies appeared to be extremely young. It is possible, scientists said, that water could be 300 to 1,300 feet below the surface. Some scientists have been skeptical of these claims, but all agree that the only way to find out is to send additional missions to Mars. The possibility of liquid water raises all types of interesting possibilities, ranging from life being found there to humans using it to sustain a permanent colony.

These robot explorers have offered a new understanding of our solar system. But the future is even more enticing, and already it is coming to fruition. On January 3, 2004, the latest mission to Mars, the *Spirit* rover, landed at Gusev Crater, which many scientists consider a former lake. After ascertaining the rover's status, scientists directed *Spirit* to send back imagery, and on January 16, to roll off its landing platform and begin exploring the red planet. The six-wheeled vehicle's first movement covered only about ten feet and took *Spirit* thirty minutes, including repeated pauses to take pictures. Its first target was a football-sized rock scientists have dubbed "Adirondack" that rested only a

few inches from the rover. Although it had a glitch in is computer software that troubled mission scientists for a time, NASA engineers have found a way to communicate reliably with *Spirit*, and to get its computer out of a cycle of rebooting many times a day. *Spirit* went on to accomplish the first brushing of a rock on a foreign planet to remove dust and allow inspection of the rock's cleaned surface using steel bristles, and it performed reliably thereafter.

Spirit's twin, *Opportunity*, landed on January 25, on the opposite of the planet from its twin, and it began geological research in Meridiani Planum, a smooth plain near the equator. Scientists found that the terrain encountered by *Opportunity* is darker than at any previous Mars landing site and has the first accessible bedrock outcropping ever seen on Mars. The outcropping immediately became a candidate target for the rover to examine up close for clues to past water activity on Mars. In a stunning discovery, *Opportunity* found evidence in early March 2004 that the red planet was once wet enough for life to exist there, although the robot had not found any direct traces of living organisms.

In this panoramic view of a Martian crater, above, Opportunity *highlights a rock outcrop near* Opportunity's *landing site. Right: An* Opportunity *image shows an extreme close-up of the rock outcrop dubbed "El Capitan," composed of layers that suggest water once flowed in abundance on the Martian surface. Below: Here* Opportunity *shows a close-up of El Capitan, which contains what appear to be rivulets where water once flowed. Overleaf: The interior of the crater in which* Opportunity *landed at Meridiani Planum on Mars.*

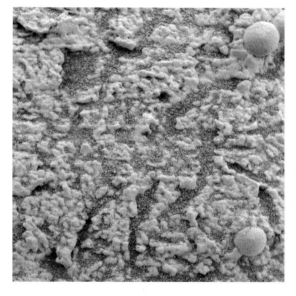

SEARCHING FOR A NEW HOME

Today, artists offer conceptual presentations of the possibilities for future space exploration. This artist's concept depicts a possible human base on Mars sometime after 2030. Created from technical information from NASA's Mars reference mission, this image depicts a two-story lander habitat, an inflatable laboratory, and a rover.

ASTRONAUT GENE CERNAN, commander of the last lunar landing mission, *Apollo 17*, in December 1972, remarked that "Mine would be man's last footstep on the Moon for too many years to come." More than thirty years have passed since Cernan stood on the Moon and expressed those bittersweet feelings about the end of an epochal event in human history. Calypso poet and troubadour Jimmy Buffett writes in "Beachhouse on the Moon" about the "relics from Apollo trips when the Earthmen came to play" as if voyages to other celestial bodies are a thing of the past. Poet B. J. van Look's "On Apollo, Before I Was Born" criticized the end of the aggressive space exploration efforts that took Americans to the Moon:

> We once had people on the Moon, before I was even born
> A year before I'd even been conceived.
> And unless I'd seen the pictures of Buzz Aldrin on the Moon
> It's nothing I'd ever have believed.

More than half of the world's population has been born since the last lunar landings, and many feel it is past the time to go back, this time to stay. They say it is also time to move outward to Mars and become a truly multiplanetary species. And we can. We have the technology, the economy, and the scientific understanding. All we need is the will. Although we have repeatedly shied away from making the commitment that would unleash the resources necessary to return to the Moon and go on to Mars, moving outward and creating a permanent human presence on these heavenly bodies represents a destiny that we cannot deny. Just after the turn of the twentieth century, Russian spaceflight philosopher Konstantin Tsiolkovskiy wrote, "The earth is the cradle of reason, but one cannot live in a cradle forever." He argued, as many have since, that the Moon and Mars represent potential homes for humanity as we move outward from Earth.

In 1989 President George H. W. Bush challenged the American public to go "back to the Moon and on to Mars," but the effort was largely stillborn because of both the lack of the infrastructure to make it possible and the detailed knowledge necessary to undertake long-term deep-space missions. In 2004, his son President George W. Bush echoed the call for a return to the Moon as a means to reaching

Mars. NASA is on its way to filling these gaps and the possibility of establishing a lunar base is now very real. NASA may also undertake a human mission to Mars, and plans have been developed toward that end.

A Passion for Discovery

One may ask, why go to the Moon and Mars? From Captain James T. Kirk's monologue — "Space, the final frontier" — at the beginning of each *Star Trek* episode to President John F. Kennedy's 1962 speech about setting sail on "this new ocean" of space, many view exploration as central to who we are and what makes us human. Astronaut and Senator John Glenn captured some of this tenor in 1983 when he summoned images of the American heritage of pioneering and argued that the next great frontier challenge was in space: "It represents the modern frontier for national adventure. Our spirit as a nation is reflected in our willingness to explore the unknown for the benefit of all humanity, and space is a prime medium in which to test our mettle."

Actor Tom Hanks, fresh from his role as Jim Lovell in the movie *Apollo 13*, made a similar case before the U.S. Congress in 1995. In a performance befitting this two-time Academy Award winner, he said:

> Feats such as going to the Moon, orbiting the Earth for weeks at a time or installing and repairing the instruments that expand our knowledge, are all celebrations of everything we Americans are supposed to be. When we decide to do so, we solve problems. We figure things out. We go into space. I know such concepts as a permanently manned orbiting science station and other NASA programs are not as glamorous as going to the Moon. And Lord knows that our problems here on our world need our attention, resolve, and service. But to choose not to go into space, to decide that our days of discovery and conquest there are over, to cease or curtail funding for the one American program that exists solely to advance the horizon for all mankind would be, I think, equal to

limiting the grand power of pure inspiration, hampering our manifest destiny, and taking away the best part of all of us.

These words represent an eloquent and moving defense of America's human space program in all its permutations.

In the last decade of the twentieth century, Americans have gained the critical elements necessary to make interplanetary voyages much more realistic than ever before: a space station served by the Space Shuttle. The Space Shuttle first began flight in 1981 as a vehicle intended to make it possible to reach low-Earth orbit with relative ease. A magnificent and complex machine — with more than 200,000 separate components that must work in synchronization with each other and to specifications more exacting than any other technological system in human history — the Space Shuttle must be viewed as a triumph of engineering and excellence in technological management. Also enormously flexible, the shuttle can carry a diversity of payloads, accomplish a myriad of tasks in orbit, and deploy and retrieve satellites, attributes that need to be considered in any effort to develop a follow-on system. A successor to the Space Shuttle must approach the same level of flexibility that this vehicle has demonstrated.

Tragedy Stalls the Program

The loss of *Columbia* on February 1, 2003, during reentry raised the specter of the viability of the Space Shuttle as an appropriate technology for human spaceflight. As the twenty-first century progresses, the fleet will likely be retired and replaced with a new human launch vehicle. Whereas NASA planners had long assumed the shuttle would be replaced in the early part of this century, efforts to develop replacements have foundered over technical or budgetary hurdles. The question of extending the Space Shuttle's service life is a major area of debate, and the decisions taken in the next few years may well set the course for the next thirty. It remains far from certain that NASA will be able to rejuvenate the aging fleet and continue its operation in a way that satisfies

These artist's concept scenes depict a possible research station on the Moon. Clockwise from right: an inflatable habitation module, a water transfer vehicle bringing water from the lunar poles, an ice mining vehicle, a small rocket for cargo transport between sites on the lunar surface, a Moon base overlook, and the inflated habitation structure.

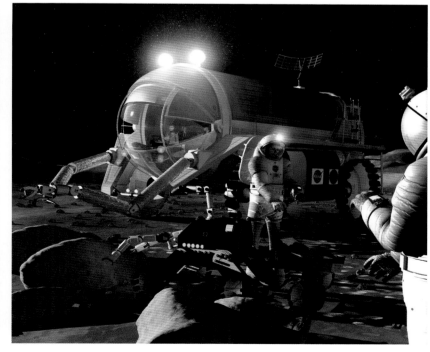

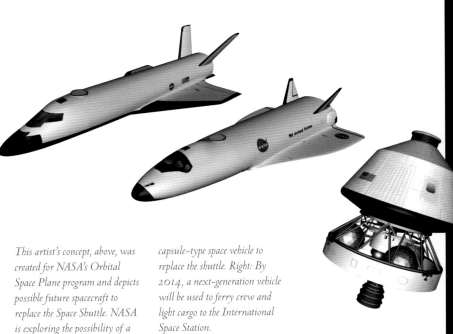

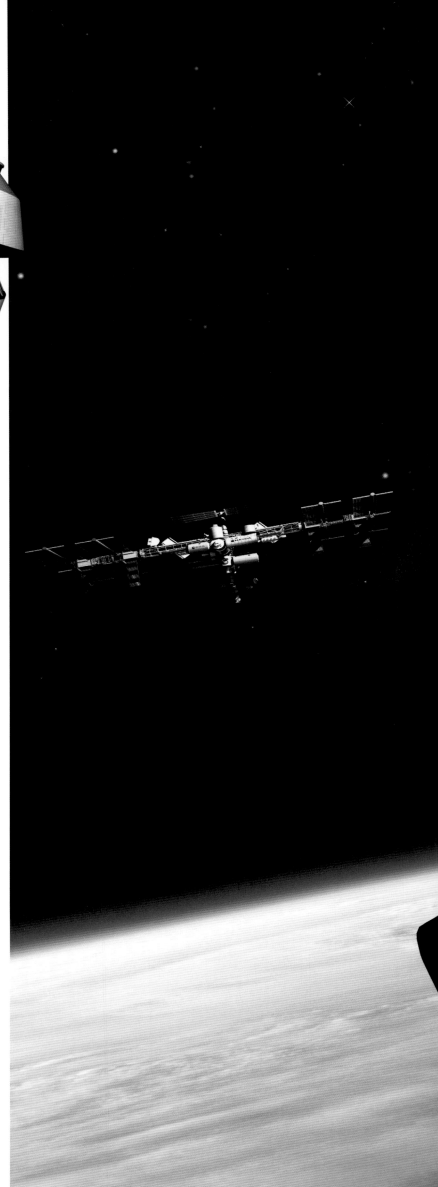

This artist's concept, above, was created for NASA's Orbital Space Plane program and depicts possible future spacecraft to replace the Space Shuttle. NASA is exploring the possibility of a capsule-type space vehicle to replace the shuttle. Right: By 2014, a next-generation vehicle will be used to ferry crew and light cargo to the International Space Station.

critics. The shuttles — the only reusable spacecraft operating in the world — were difficult enough to maintain when their thousands of parts were new, but the effects of aging have added significantly to the challenge. All of this has brought an immeasurable degree of uncertainty about the future. There is good reason to believe, however, that NASA's Orbital Space Plane (OSP) program, a replacement for the Space Shuttle currently in the works, will yield a next-generation launcher capable of supporting human missions into Earth orbit.

From the International Space Station (ISS), served first by the shuttle and later by an OSP vehicle, humans may well launch missions back to the Moon and on to Mars. The creation of a permanently occupied space station, something that has long been a critical component in space architecture, is finding realization in the ISS. Although sidetracked by the Apollo program, in the post–Cold War era, the United States and the former Soviet Union have joined with fourteen other nations to make a reality the long-held vision of a space station in Earth orbit. This relationship made the ISS a reality in 2000 when the first crew set up resi-

dence aboard the craft. With this accomplishment, the spacefaring nations of the world intend that no future generation will ever know a time when there is not some human presence in space. Once functioning in space, the station should energize the development of private orbital laboratories. Such laboratories would travel in paths near the ISS. The high-tech tenants of this orbital "research park" would take advantage of the unique features of microgravity. This permits research not possible on Earth in such areas as materials science, fluid physics, combustion science, and biotechnology.

A FEASIBLE LUNAR GOAL

Using this space station as a base camp, humans will return to the Moon and establish a permanent human presence there. Because the necessary technology for getting there already exists in the Space Shuttle and the ISS, all that is required is a lunar lander hoisted to Earth orbit inside the payload bay of the shuttle and launched from the ISS. Much like the lunar module of Apollo, this vehicle may look quite ungainly, but it operates only beyond Earth's atmosphere and gravity. Such an endeavor will require only a moderate investment, and the results may well be astounding.

Bridging the Present and Future

Sally Kristen Ride, the first American woman to fly in space, has been significant in ensuring the safety of human spaceflight since the 1980s. Born on May 26, 1951, in Los Angeles, California, Ride began early to combine her competitive spirit with academic determination. As a youngster, her ability on the tennis courts led her to rate eighteenth nationally on the junior tennis circuit. Initially studying physics at Swarthmore College, she finished at Stanford University, earning her B.S. in physics and B.A. in English literature in 1973. Ride received her Ph.D in physics at Stanford University in 1978.

Hearing that NASA wanted young scientists to serve as shuttle mission specialists, she applied and was selected in the astronaut class of 1978. She made her first and only flight in 1983 aboard STS-7, the flight that set the pattern for combined-gender shuttle flights.

After the *Challenger* accident in January 1986, Ride served on the Presidential Commission on the Space Shuttle Challenger Accident in 1986. In 1986–87 she chaired a NASA task force that prepared a report on the future of the civilian space program. Entitled *Leadership and America's Future in Space,* it served as a blueprint for restructuring the agency after the accident and helping it return to space flight.

After leaving NASA in 1987, Ride first joined the Center for International Security and Arms Control at Stanford University, and in 1989 she assumed her current position as director of the California Space Institute, part of the University of California at San Diego.

With the space shuttle *Columbia*'s tragic loss on February 1, 2003, Ride served as a member of the Columbia Accident Investigation Board and contributed important insights into the deliberations relating to crew safety.

Astronaut Sally K. Ride in 1983

With accessibility to the Moon relatively routine, building a human outpost there is within reach. And unlike Earth orbit, the Moon has an abundance of materials available from which to create a self-sufficient base. This will probably be established just below the surface near the Moon's poles, where ice has been detected. Using ice from the poles, humans will be able to create water, oxygen, and hydrogen. Using solar energy may also create virtually unlimited electrical power. These are critical components for a permanent human presence, and they already reside on the Moon in abundance.

COOPERATION NEEDED

Perhaps the hardest part of spaceflight is not the scientific and technological challenges of operating in an excep-tionally foreign and hostile environment but in the environment of rough-and-tumble international and domestic politics. For all of its possibilities, one must also recognized that international cooperation in space exploration is an enormously difficult process. Rocket pioneer and space exploration advocate Wernher von Braun once said, "We can lick gravity, but sometimes the paperwork is overwhelming." Even so, cooperative space endeavors have been richly rewarding and overwhelmingly useful, from all manner of scientific, technical, social, and political perspectives. A lunar base as a cooperative venture could also help the people on Earth to live together in greater harmony. The Moon in the next few years, like Antarctica, will become an international protectorate with scientists from many nations of the Earth permanently in residence. In the process, more will be learned about the universe and humanity.

While many hold a romantic notion that humans will colonize the Moon, undertaking economic activities there that will drive immigration and eventually tourism, those possibilities are still far in the future. It will probably be the middle part of the twenty-first century before Moon bases will move beyond a rudimentary stage and permanent colonies become the norm.

POTENTIAL HOSPITALITY

No long-lived technological civilization can remain forever on its home planet and survive, astronomer Carl Sagan proclaimed in his book *Pale Blue Dot*. The Sun will eventually die; an asteroid will batter civilization; or some other catastrophe will occur. The eventual choice, Sagan said, "is spaceflight or extinction." Humans became the dominant species on Earth by their willingness to migrate out of Africa and settle terrestrial regions where they could not live without the accruements of technology. Sagan fully expected that the basic human drive to settle new lands would continue in space.

The logical place beyond the Moon for human colonization is Mars. So rich and inviting, Mars captures our vision of an expansive human presence beyond Earth that allows us to become a truly multi-planetary species. More than one hundred years ago, a Boston Brahmin astronomer named Percival Lowell captured the imagination of people

Artists have depicted the Moon as a possible future abode of life from Earth. Left: Firing advanced cryogenic fuel engines, a reusable Lunar Lander leaves an established lunar outpost. Top: An antenna installer has fallen and is transported to the lunar base. Above: A Moon base could test many relevant Mars systems and technologies. Right: A lunar outpost crew prepares wiring for a communications relay.

throughout the world by describing more than two hundred canals that appeared to link polar ice caps to regions near the Martian equator. These canals, said Lowell, provided evidence that complex beings had evolved on a planet not entirely hospitable to life and that they had survived by transferring increasingly scarce water from polar regions to warmer zones only through global cooperation. These speculations gave rise to the myth of Mars, a motivating force in human curiosity about the planet. People genuinely expected to find life there. While robotic exploration of the planet dashed all hope of finding complex life on the surface and proved the canals a figment of Lowell's imagination, data from *Mars Global Surveyor* and other probes offered tantalizing possibilities that

water, and therefore life, may have once been abundant on Mars.

These findings have energized a major effort to send humans to Mars. The Mars Society, founded by quintessential Mars advocate Robert Zubrin, has argued that using a "Mars Direct" approach, Earth could send a small mission to the planet by "living off the land." If Meriwether Lewis and William Clark had been forced to carry all their fuel and supplies across the continent, he insists, they would have been stuck forever in Missouri. In a similar fashion, Zubrin insists that the first humans to Mars should extract fuel and consumables from the Martian environment. Zubrin and other advocates of Mars Direct, including some inside NASA, insist that a human expedition to Mars could be conducted for about $50 bil-

lion. Such an undertaking would require a Mars exploration program funded at about $6 billion per year for eight years, a diminutive sum given the size of the U.S. economy. According to Zubrin, the first landing would occur within eight years of funding and landings would continue every two years thereafter.

To accomplish this mission, six years after funding began engineers would launch to Mars three large rockets from the Kennedy Space Center in Florida. Each vehicle would resemble a NASA space shuttle—two large boosters attached to a 27.5-foot-wide external fuel tank supplemented by hydrogen-burning engines on the aft end. In place of the familiar orbiter, however, the rockets would support a cylindrical object containing 176,000 pounds of payload. The

first rocket would contain a habitat module, a rocket motor, and consumables for the first crew. It would go into orbit around Mars, awaiting a homebound crew. The second vehicle would land a habitat module and other support equipment on the surface of Mars. The third craft would deliver more equipment to the Martian surface, including an ascent vehicle for departing Mars and a manufacturing plant to produce propellant from chemicals in the atmosphere.

After the three spacecraft successfully prepared the way, including manufacturing the fuel for the return trip, a six-astronaut expedition would begin a twenty-six-month mission to Mars. Upon arrival, the crew would spend six hundred days exploring the planet. They would collect geological samples for analysis in their laboratory,

When humans go to Mars, it may look very much as depicted in these images created for NASA from technical descriptions. *Above:* A cargo lander unloads on the Martian surface. *Far left:* The crew's ascent vehicle and propellant production facility can be seen one kilometer away from the completed outpost. *Near left:* The crew attaches an inflatable laboratory to their lander to increase the internal pressurized volume of their Martian home.

and seek evidence of water and life both past and present. Most important, they would seek natural resources that would be useful for future missions to the red planet. Returning to Earth could take as much as 110 days, and upon arrival home they could either rendezvous with the International Space Station or splashdown on Earth "Apollo style."

Advocates of this approach to Mars exploration emphasize that a long series of these missions could take place over many years, each building on the last and eventually creating the infrastructure on the planet for a permanent human presence. When a second crew arrived, for instance, they would find already in place the manufacturing plants and habitat modules from the last expedition at the landing zone, along with return vehicles and an infrastructure of satellites and ground stations. Using this methodology,

humans could establish the first research station on Mars.

OBSTACLES TO A MARTIAN COLONY

Even so, the human exploration of Mars is a daunting task. A majority of Americans do not support human missions to Mars and never have. Consistently, only about 40 percent of those polled have supported human missions to Mars. In that climate there is little political justification to support an effort to colonize the planet. No less challenging are the technical and scientific challenges of sending humans so far through such a hostile environment and recovering them successfully. Just protection from radiation for the crews alone presents enormous problems over such a long period.

Of course, the United States could send human expeditions to Mars. There is nothing magical about it, and a national mobilization to do so could be successful. But a human

Mars landing would require a decision to accept enormous risk for a bold effort and to expend considerable funds in its accomplishment over a long period. Using Apollo as a model — addressed as it was to a very specific political crisis relating to U.S.–Soviet competition — anyone seeking a decision to mount a human expedition to Mars must ask a critical question: What political, military, social, economic, cultural challenge, scenario, or emergency can one envision to which the best response would be a national commitment on the part of the president and other elected officials to send humans to Mars? In addition, with significantly more failures than successes, and half of the eight probes of the 1990s ending in failure, any mission to Mars is greater in complexity, risk, and cost than returning to the Moon. Absent a major external factor that would change the space policy and political

For several years, the Mars Society has worked to study extreme environments on Earth that might help humanity better understand how to survive on Mars. In 2001 the Society created the Mars Desert Research Station in the barren canyonlands of Utah to simulate living and working on another planet. The Station's centerpiece is a cylindrical habitat, upper far right, an eight-meter-diameter two-deck structure mounted on landing struts very much like those proposed for use on Mars. A team of *up to six crew members live for months at a time in relative isolation in this Mars analog environment, two of whom are shown in 2002 examining rock samples during an EVA complete with spacesuits, near right. Lower far right, in February 2002 crew members Heather Chluda, Troy Wegman, and Jennifer Heldmann used four-wheel all-terrain vehicles to explore the region between the local ridge near the habitat and Skyline Rim.*

Imagining the Way

Sir Arthur C. Clarke has inspired generations of space enthusiasts through both his science fiction and nonfiction writings about space exploration. Born in Great Britain in 1917, Clarke burst onto the world stage in 1945 with his technical paper "Extra-terrestrial Relays," in *Wireless World,* setting out the principles of satellite communication in geostationary orbit at approximately 26,000 miles about the Earth. Geostationary orbit is now officially named "The Clarke Orbit" by the International Astronomical Union in honor of his contribution. But Clarke is even better known for his powerful science fiction. *Childhood's End* (1953) tells the story of Overlords helping humanity to eliminate ignorance, disease, and poverty, and preparing them for the next step in evolution. *2001: A Space Odyssey* (1968) was the culmination of four years of collaboration with film director Stanley Kubrick on one of the most visually stunning representations of human exploration of the solar system ever undertaken. The film, which earned for Clarke and Kubrick an Oscar nomination, has profoundly influenced virtually all spaceflight advocates since that time. Clarke lives in Sri Lanka, his home since 1956, and continues to write and speak via telecommunications satellite to the world. He has developed three laws that ring true, especially during the space age:

■ When a distinguished but elderly scientist states that something is possible, he is almost certainly right. When he states that something is impossible, he is very probably wrong.

■ The only way of discovering the limits of the possible is by venturing a little way past them into the impossible.

■ Any sufficiently advanced technology is indistinguishable from magic.

Sir Arthur C. Clarke, advocate of space exploration

landscapes, humans will probably not land on Mars before 2030.

This is not a cause for despair, for it suggests an orderly stepping-stone approach to becoming a multiplanetary species. First humans will return to the Moon, this time to stay. From that foothold beyond Earth, we could well move on to Mars. Perhaps we will learn in the next quarter century what we need to know to make human visits to Mars a reality. For instance, humans need to learn how to keep strong during an extended weightless/low-gravity environment for more than seven hundred days for a complete Mars mission. The debilitating effects of permanent bone loss and vascular system damage during only three hundred days aboard the Mir space station suggest that more research, and the development of better countermeasures, are necessary before mounting a Mars expedition. No less significant, the effects on astronauts of extended exposure to the radiation of deep space present significant challenges that must be mastered prior to beginning such a mission.

WHAT TO EXPECT

If these plans are right, humans could plan for an expedition to Mars in 2035, when the planet will be in "opposition." Its closest point to Earth and the Sun's periodic ebb and flow of solar flare and sunspot activity will be at its lowest point, thereby minimizing solar radiation and cosmic ray exposure, during the mid-2030s. The next time in the twenty-

first century when these events will coincide is 2065.

In a stunning development on January 14, 2004, President George W. Bush announced that he was giving NASA the mandate to return to the Moon, this time to create a permanent human presence, between 2015 and 2020, and there to prepare for the human exploration of Mars. To accomplish this task, the president challenged NASA to refocus its energies toward these goals by committing the ISS to research necessary to fly into deep space and to energize all exploration programs to support these ultimate objectives. He commented, "Today I announce a new plan to explore space and extend a human presence across our solar system. We will begin the effort quickly, using existing programs and personnel. We'll make steady progress—one mission, one voyage, one landing at a time." Upon reaching the Moon, he added, "with the experience and knowledge gained on the moon, we will then be ready to take the next steps of space exploration: human missions to Mars and to worlds beyond." It remains to be seen if humanity will seize upon this mandate to occupy the Moon and move on to Mars. It is incumbent on the policy makers and the public to make decisions about sustained exploration. Will the twenty-first century realize the promise of the first space explorations of the first lunar landings?

Artist conceptions depict a future on Mars in which geologists and biologists can explore the planet, piecing together its history and perhaps evidence of ancient life. Below: Sojourner is visited many years after its mission by a researcher. Bottom: Two scientists seek fossils that would demonstrate evidence of past life on the Martian surface. Opposite: A geologist examines fossils preserved in the rocks.

THE FINAL FRONTIER

OUR AWESOME UNIVERSE

ONLY FIVE CENTURIES AGO, not even the blink of an eye in time when compared to the age of the universe, humanity's vision extended but a little beyond Saturn. Our ancestors envisioned a universe both limited and orderly. The telescope changed all that, and our universe expanded exponentially as we observed literally thousands of objects beyond Earth. The most significant astronomical instrument since Galileo's first telescope, the Hubble Space Telescope, has also revolutionized our understanding. In just the last decade, our cosmology has changed as we found evidence of black holes, the Big Bang, dark matter, dark energy, the corpuscular universe, multiple dimensions, and extrasolar planets. So just what have we learned since the beginning of the space age, what big questions exist at present, and how are space scientists seeking to answer them?

Using the powerful telescope at the Mount Wilson Observatory near Pasadena, California, in the early 1920s, astronomer Edwin Hubble first confirmed the existence of galaxies outside the Milky Way. He then observed that these galaxies were racing away from ours, an indication that the universe was expanding. This became the first critical evidence pointing to an explanation of the origin of the universe in a Big Bang from a singularity of infinite nothingness that contained all of the matter of our universe to what we observe today. It proved an elegant, convincing, and resilient theory of the universe's formation. Despite periodic detractors, the Big Bang theory is virtually universally accepted in the scientific community even as it undergoes modification through further observation and experimentation.

For decades scientists debated the possibility that the mass of the universe—matter, anti-matter, dark matter, and perhaps others types as yet undiscovered—would be sufficient that its gravitational forces over eons would halt the expansion and begin a contraction of the universe back to its singularity. And then the process would start all over again. Whether we live in an open (expanding indefinitely) or closed (eventually contracting) universe depends very much on the mass of the universe and the gravity generated by it. Thus far scientists have found that only about 4 percent of the universe exists in the form

The Wilkinson Microwave Anisotropy Probe (WMAP), above, is a small but powerful spacecraft seeking the location of the Big Bang. Right: The Microwave Sky is a detailed, all-sky picture of the infant universe produced by WMAP. This image reveals thirteen-billion-year-old temperature fluctuations (shown as color differences) that correspond to the seeds that grew to become the galaxies. Encoded in the patterns are the answers to many age-old questions, such as the age and geometry of the universe.

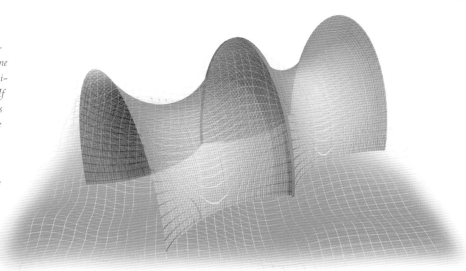

of stars, gas planets, white dwarfs, black holes, and other observable matter. The rest is made up of dark energy or dark matter, and the core question is whether this 96 percent of the universe's "stuff" is sufficiently strong in gravity to pull everything back from endless expansion. Many scientists now think that the universe will continue to expand indefinitely, with energy eventually dissipating some trillion years in the future and all stars eventually becoming just so many stellar cinders. This is a pretty bleak future, but obviously a far distant one.

More than One Universe

Of course, the universe may not end in this manner. Scientists may be wrong about these projections. Some yet undiscovered mass or process may affect cosmic evolution in a far different manner. The most intriguing possibility at present is a theory of multiple universes, referred to as multiverses, occupying the same space-time. According to some cosmologists, the space in our universe is part of a three-dimensional place called a brane, short for membrane, which is also part of a much larger eleven-dimensional space-time connected by gravitational forces. There may even be an infinite number of three-dimensional universes occupying portions of this larger eleven-dimensional space-time. In modern M-theory, as the multiverse ideas are called, other universes might be very different from ours, but separate from own space-time by a very short physical space. Like frequencies on a single radio, there might

be a vast number of universes co-existing in an eternal state.

Perhaps black holes offer linkages between these universes. Perhaps there are ways to travel between them. Perhaps, also, black holes are the point from which new universes begin, extending as tentacles of the universe from which the black hole had emerged. This theory is unproven, and perhaps unprovable, but it represents an enormously exciting potential for future research.

Less strange but no less extraordinary, in just the last two decades three enormously exciting areas of research have yielded stunning scientific results. The first of these involves using the Hubble Space Telescope, the Cosmic Background Explorer (COBE), the Chandra X-ray Observatory, and other satellites to peer back to the Big Bang in search of how the universe began. Already they have determined the age of the universe with great precision. On February 11, 2003, NASA astronomer Charles Bennett announced that using the Wilkinson Microwave Anisotropy Probe (WMAP) his team had found that the universe is 13.7 billion years old, with a margin for error of a miniscule 100 million years. Furthermore, WMAP data suggest that our Milky Way was formed within 200 million years of the Big Bang. Our Sun and its solar system, however, is much younger, formed only about 4.51 billion years ago.

A Galaxy Is Born

Since 1990, the Hubble Space Telescope has captured the birth of

stars and galaxies. Peering back more than 10.5 billion light-years, Hubble captures light dating from three billion years after the birth of the Milky Way. These primordial galaxies tell scientists much about the origins of our own star and galaxy. One photograph from 1995, for instance, captured galaxies that may have formed soon after the universe began. The so-called "Pillars of Creation" image is one of the most striking taken by Hubble, for it captured the nursery of newborn stars forming within gaseous globules in the Eagle Nebula in another part of our own Milky Way galaxy seven thousand light-years away. The stars depicted in the image are so far away that the light from them has taken almost the age of the universe to reach the Earth.

Equally spectacular, in 1989 NASA launched the Cosmic Background Explorer to measure the temperature of space in an attempt to determine the place where the Big Bang took place. COBE captured variations in the temperature of what scientists call background radiation, remnants of the moment when the universe emerged from a dense fog of opaque light and became transparent. Pinpointing that location, COBE produced an image of the universe when it was only 300,000 years old. The instrument therefore revealed the seeds from which the first galaxies formed.

This search for the location of the Big Bang and its attendant birth of galaxies and stars leads directly to the second exciting area of research in space: the discovery of extrasolar planets, especially Earthlike ones.

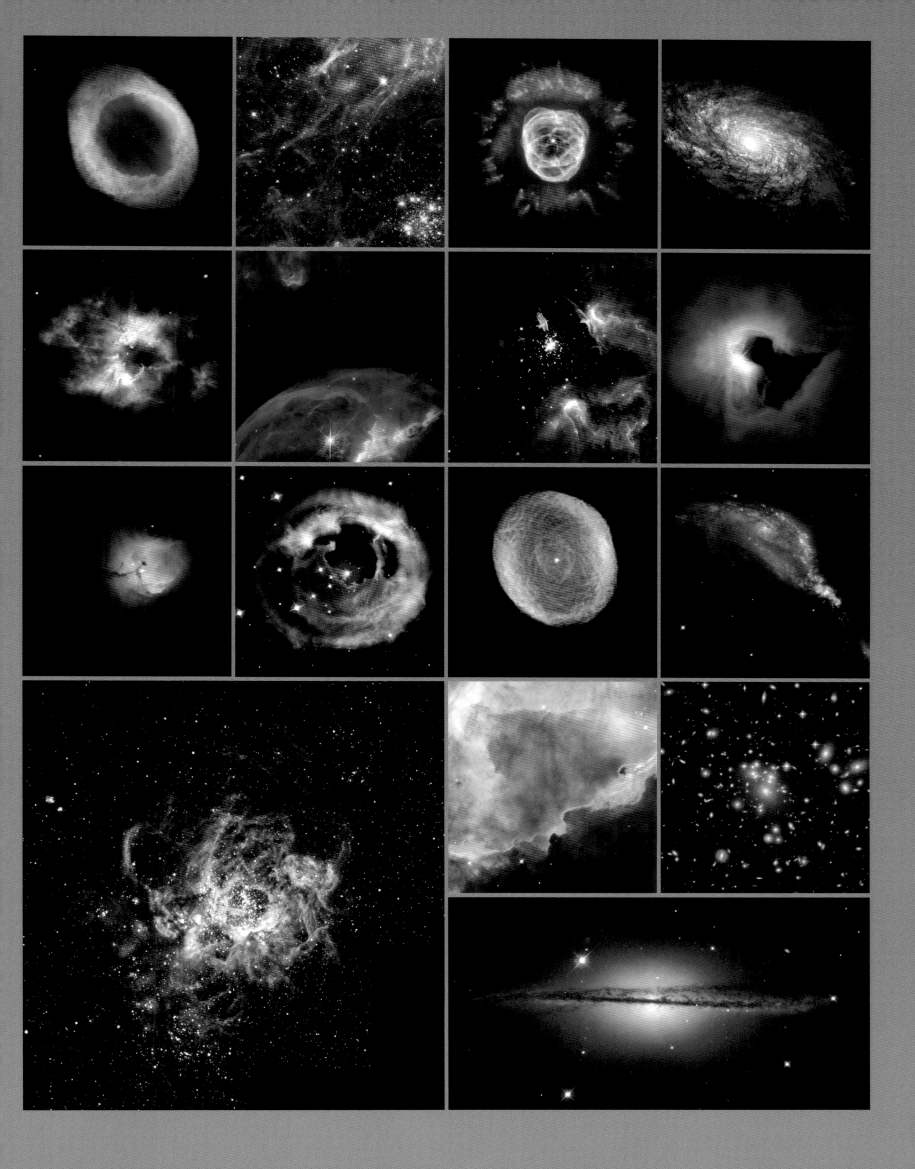

Explaining the Universe

Many consider British cosmologist Stephen Hawking the most brilliant theoretical physicist since Albert Einstein. He has made seminal contributions in the study of black holes, particle physics, supersymmetry, space-time, and quantum gravity. Hawking's thinking on the origins and the end of the universe are critical in modern understanding of the subjects. Born in 1942 and educated at Oxford and Cambridge Universities, Hawking has long sought a Grand Unification Theory (GUT) that would link Einstein's theory of relativity, which deals with gravity, with quantum mechanics, which deals with events at an atomic and subatomic level.

Hawking's work on space-time suggests that perhaps there are multiverses which give birth to follow-on universes one after another. Best known to the public for his highly accessible *A Brief History of Time: From the Big Bang to Black Holes* (1988), Hawking laid out his GUT as well as the evolution of his thinking about the origins, evolution, and death of the universe and its place in the multiverses he believes exist in space-time. His work may have been slowed by his chronic illness, amyotrophic lateral sclerosis (ALS), sometimes called Lou Gehrig's disease, which confines him to a wheelchair and requires that he use a computer with a voice synthesizer to speak. Others believe that ALS spurred his creativity. Regardless, Stephen Hawking is one of the most original and pathbreaking thinkers in science.

Stephen Hawking, perhaps the most significant cosmologist since Albert Einstein

Scientists have believed throughout the twentieth century that planets exist around many stars, even ones located in globular star clusters where Earth-like elements do not abound. Within the last decade, scientists have used a number of techniques to find planets beyond this solar system. These include pulsar timing, which measures the pulses indicating an orbiting object around another star; transit photometry, which measures the reduction of light from a star as a planet passes in front of it; spectroscopy, which measures slight changes in the velocity of the star as a planet acts on it; and gravitational lensing, which monitors the changes in the gravitational field.

All of these, of course, are indirect methods of finding extrasolar planets, but they have been quite successful at detecting large planets the size of Jupiter. Using these methods, as of June 2003 astronomers had found 108 extrasolar planets in orbit around 94 normal stars and two pulsars. And we know quite a lot about some of these planets. For example, planet HD 209458b has a mass that is 0.7 of Jupiter's, is 1.3 times the size of Jupiter, and is traveling in an orbit around its star at of four million miles. Using the Hubble Space Telescope and other instruments, scientists have determined that this planet is so hot and so close to its star that its hydrogen is boiling off and forming a comet-like trail.

In the first part of the twenty-first century scientists will deploy new space-based instruments, especially interferometers and spectrometers, to find planets with ever more precision

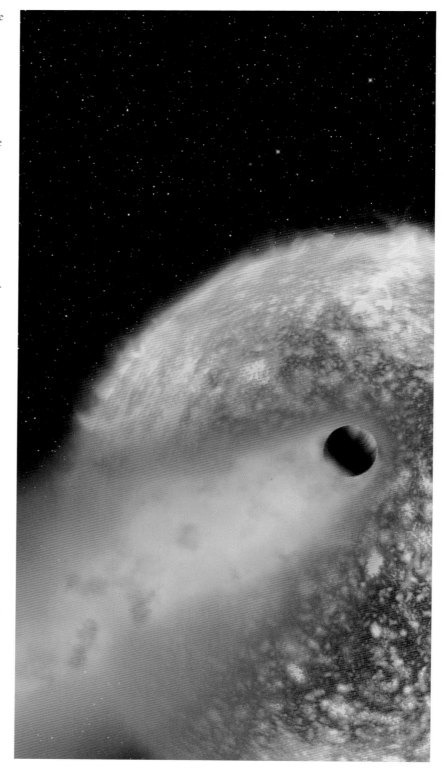

This artist's illustration, left, shows a dramatic close-up of the scorched extrasolar planet HD 209458b in its orbit only four million miles from its yellow, Sun-like star. Top: When Apollo 12 astronaut Pete Conrad brought back from the Moon components from Surveyor 3 in 1969, scientific analysis determined that microbial life forms from Earth infesting the spacecraft had survived on the lunar surface. Above: In exploring the depths of the ocean, scientists have found life surviving near sea vents like these, which provide heat and energy.

and detail. Perhaps scientists will be able to observe Earth-like planets that could become habitats for humanity. NASA's Kepler spacecraft, set for launch in 2006, will monitor the brightness of more than 100,000 stars with the goal of discovering planets orbiting in a habitable zone around a star. The best estimates are that this mission will discover more than nine hundred Earthlike planets around other stars; all future targets for interstellar exploration. Kepler is just the first in a series of missions aimed at finding terrestrial planets. NASA scientists anticipate that the Terrestrial Planet Finder mission, to be launched after 2010, will continue Kepler's work and find even more Earthlike planets.

PLACES LIKE EARTH

The discovery of extrasolar planets has lain to rest any thought that our solar system is the only one in existence in the universe. The discovery of Earthlike planets circling other stars will go far toward proving the third great area of exploration in space, the search for life beyond this planet. The possibility of finding concrete evidence of life elsewhere in the universe this century is very great.

Meanwhile, scientists are trying to understand the potential for life developing elsewhere. In 2001, astrobiologists Lou Allamandola and Jason Dworkin demonstrated at NASA's Ames Research Center in Mountain View, California, that they could mix common elements found in stellar clouds and create cell membranes. With further steps replicating the evolution of planetary atmospheres, amino acids and other building blocks of DNA can also be created in a laboratory. The ease with which these steps have yielded prebiological molecules suggests that life may be abundant in the universe. Moreover, the discovery during the 1990s of myriad new sea species by biologists in the depths of our oceans near sea vents spewing superheated gases from the Earth's core suggests that where there is both water and any type of energy source, life will exist.

Despite the fact that no one produced direct, irrefutable proof of extraterrestrial life during the twentieth century—a matter of considerable disappointment to people engaged in the search—most scientists believe that some type of simple life, perhaps virus or bacteria, exists in the solar system. We know that simple life, once it begins, is remarkably robust. Streptococcus bacteria from Earth survived inside the television camera of the *Surveyor 3* space probe after it landed on the Moon in April 1967. The bacteria survived nearly three years of radiation and average temperatures only 20 degrees above absolute zero. *Apollo 12* astronauts brought the camera back to Earth at the end of 1969, where the bacteria regenerated themselves. Moreover, experience on Earth suggests that microbial life appears as soon as suitable conditions arise. The range of conditions is remarkably wide. Bacteria-like organisms exist in high-temperature geysers. They live miles below the surface of Earth in subterranean darkness. The frequency with which life begins is an important component in the formula for calculating the possible number of extraterrestrial homes. Scientists believe that life begins with remarkable ease, and they are seeking possible locales in the solar system conducive to this type of simple life. Mars's substrata might be one; an ice-covered ocean on Europa, a moon of Jupiter, might be another.

Sometime within the first two decades of the twenty-first century, NASA will undertake a sample return mission from Mars. The vehicle will land in an area of possible biological activity, such as an ancient lakebed. Its rover will travel ten or more miles from the landing site in search of samples. While the rover is out collecting samples, the return vehicle will collect gases from the Martian atmosphere and, utilizing a small on-board chemical plant, manufacture rocket propellants—

probably methane and oxygen. The rover, on returning, will transfer its samples to sturdy containers on the return vehicle, which will then return to Earth. Life on Earth and Mars may have had a common origin, and if so we want to know about it. Sample return missions will help us learn much more by comparing genetic codes. Perhaps life began on Mars, invaded Earth, and flourished. All earthly life may be Martian in origin—or vice versa. The possibilities are breathtaking.

Also a possibility for the not-too-distant future is a mission to Europa. Scientists hope to find simple bacteria in the oceans under the ice of Europa; but they will also look for more complex forms such as the tubeworms that exist along Earth's ocean floor. To search for life, scientists will send submersible robots to explore Europa's subterranean sea. This will be technically challenging, since the icy crust could be as much as one hundred miles thick. Orbiting spacecraft scanning the surface of Europa will search for thin spots in the ice and dispatch a robotic Lander to the surface. There, a cryobot—a robot capable of operating in icy cold or cryogenic conditions—will melt through the ice, trailing a communication line behind. Once below the icy crust, the cryobot will release a submersible hydrobot—a robot designed to work under the sea—which explores the seabed and any creatures living on it. A camera and a variety of instruments will survey the undersea environment and communicate the findings back to Earth.

Is There Life Out There?

Although the most immediate research will look for simple forms of life, many people also believe that more sophisticated life forms probably exist elsewhere in the universe. More than one-third of the American public believes that creatures "somewhat like ourselves" reside on other planets, a belief reinforced by a plethora of films and novels anticipating the shape of alien beings. They may be right; however, there is not one scintilla of evidence to support this assertion as yet.

In a 1975 *Scientific American* article, astronomers Carl Sagan and Frank Drake estimated there may be more than one million civilizations in the Milky Way galaxy "at or beyond the Earth's present level of technological development." They derived their million-civilization prediction from a simple formula based on estimates of the number of solar systems in the galaxy, the number of planets suitable for life, and the fraction of suitable planets where life can evolve into complex forms.

Humans know from their own experience that complex life forms build machines. Alien creatures on other planets may be quite noisy, broadcasting signals into the void just as humans have been doing with radio and television waves for one hundred years. For many years a small band of astronomers have been scanning the heavens in search of the proverbial needle in a haystack. During the 1970s, scientists won government funding for the endeavor, which they called the Search for Extraterrestrial Intelligence (SETI),

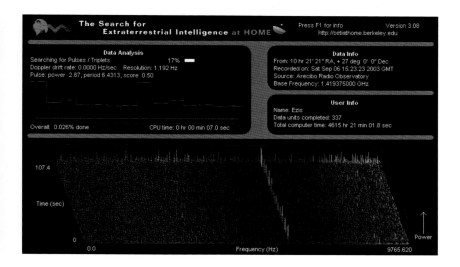

The Search for Extraterrestrial Intelligence at HOME

but in 1993 the United States government terminated all tax-supported funding. SETI advocates sought and received private funding, and the work continues through the SETI Institute, a nongovernmental body.

Collecting the Evidence

SETI scientists use telescopes to collect radio waves from nearby stars. For each observation of a single star, computers examine tens of millions of channels within a ten-megahertz bandwidth. Computers filter out signals from human sources, including the growing number of telecommunication satellites, and compare the results to the universe's natural background noise. Any strange or unfamiliar signal triggers a special procedure, called FUDD, or follow-up detection device. Two separate radio telescopes, hundreds of miles apart, track the signal and apply separate FUDDs. It is a tedious process. A signal from an alien civilization could come tomorrow; one may never come at all.

The SETI program staged a public relations coup by creating the SETI@home project, an innovative idea to involve millions of people around the world who donate their computers' extra processing capability to analyze radio signals, hoping to find signs of intelligent life elsewhere in the universe. Anyone with a computer and an Internet connection can participate by running a program that downloads and analyzes radio telescope data. It has been an enormously popular effort, for it appeals to millions who relish the small but captivating possibility that their computer will detect the faint murmur of a civilization beyond Earth. The participants receive data recorded from the radio telescope at Arecibo, Puerto Rico, and then compute a power spectrum of the radio signal, looking for interesting, unusual, and nonnatural radio signals.

The results are sent to the SETI@home project for detection confirmation. Whoever identifies a pattern that proves to be an actual signal will be named co-discoverer of extraterrestrial life.

SETI@home is probably the first of many efforts that space scientists will unveil in the years ahead. Providing Earth-bound humans the opportunity to participate in the excitement of scientific discovery will motivate space exploration increasingly as the twenty-first century unfolds. This is because it taps into the core question of all humanity: are we alone in the universe? We do not think so, but confirming that belief may be one of the most deserving objectives ever realized.

Space exploration provides a window on the universe from which fantastic new discoveries can be made. Humans may well discover extraterrestrial life. They may set their eyes on the image of an Earth-like planet around a nearby star. They may discover some fantastic material that can only be made in a gravity-free realm. Perhaps they will discover some previously unknown principle of physics. Certainly, they will capture an image of the creation.

Everything about our future in space is possible, but nothing is guaranteed. Only one feature of space exploration is inevitable: surprises. Space is full of achievements, disappointments, and surprises. By going into space, humans learn what they do not know. Ultimately, we may learn to live on a small and precious world by learning to move beyond it.

A BASIC SHELF OF READINGS ON THE HISTORY OF SPACE EXPLORATION

Baker, David. *Spaceflight and Rocketry: A Chronology.* New York: Facts on File, 1996. A basic, single-volume reference on space exploration.

Bilstein, Roger E. *Flight in America: From the Wrights to the Astronauts.* Baltimore, MD: Johns Hopkins University Press, 1984, paperback reprint 1994. A superb synthesis of the origins and development of aerospace activities in America. This is the book to start with in any investigation of air and space activities.

Burrows, William E. *The Infinite Journey: Eyewitness Accounts of NASA and the Age of Space.* New York: Discovery Books, 2000. A large-format history of space flight that includes important discussions of the endeavor by participants.

——. *This New Ocean: The Story of the First Space Age.* New York: Random House, 1998. A strong overview of the history of the space age from Sputnik to 1998.

Caiden, Martin, and Jay Barbree, with Susan Wright. *Destination Mars: In Art, Myth, and Science.* New York: Penguin Studio, 1997. A beautifully illustrated overview of the lure of the red planet throughout humanity's history.

Chaikin, Andrew. *A Man on the Moon: The Voyages of the Apollo Astronauts.* New York: Viking, 1994. One of the best books on Apollo, this work emphasizes the exploration of the Moon by the astronauts between 1968 and 1972.

Collins, Michael. *Carrying the Fire: An Astronaut's Journeys.* New York: Farrar, Straus and Giroux, 1974. This is the first candid book about life as an astronaut, written by the member of the *Apollo 11* crew that remained in orbit around the Moon.

——. *Liftoff: The Story of America's Adventure in Space.* New York: Grove Press, 1988. General history of the U.S. space program for a popular audience written by a former astronaut.

Darling, David. *Life Everywhere: The Maverick Science of Astrobiology.* New York: Basic Books, 2001. A study of life in the universe.

Dick, Steven J. *The Biological Universe: The Twentieth Century Extraterrestrial Life Debate and the Limits of Science.* New York: Cambridge University Press, 1996. The superb history of the possibility of life elsewhere in the universe.

Fischer, Daniel. *Mission Jupiter: The Spectacular Journey of the Galileo Spacecraft.* New York: Copernicus Books, 2001. A history of the *Galileo* spacecraft to Jupiter.

Jenkins, Dennis R. *Space Shuttle: The History of the National Space Transportation System, the First 100 Missions.* North Branch, MN: Speciality Press, 2001, 3rd Edition. By far the best technical history of the Space Shuttle, presenting an overview of the vehicle's development and use.

Kevles, Bettyann Holtzmann. *Almost Heaven: The Story of Women in Space.* New York: Basic Books, 2003. A fascinating account of the experiences of women in space.

Launius, Roger D. *Frontiers of Space Exploration.* Westport, CT: Greenwood Press, 2nd edition, 2003. A collection of essays along with key documents and biographies of actors in the space exploration effort.

——. *Space Stations: Base Camps to the Stars.* Washington, DC: Smithsonian Books, 2003. A history of space stations—real and imagined—as cultural icons, fully illustrated with rare and evocative imagery.

——, and Howard E. McCurdy. *Imagining Space: Achievements, Predictions, Possibilities, 1950-2050.* San Francisco, CA: Chronicle Books, 2001. A large-format book sweeping from the past to the future of spaceflight.

Lemonick, Michael D. *Echo of the Big Bang.* Princeton, NJ: Princeton University Press, 2003. An excellent discussion on the current state of cosmology and why.

——. *Other Worlds: The Search for Life in the Universe.* New York: Simon & Schuster, 1998. A strong analysis of whether or not there is life on other planets in the universe and efforts underway to find it.

Leverington, David. *New Cosmic Horizons: Space Astronomy from the V-2 to the Hubble Space Telescope.* New York: Cambridge University Press, 2001. An outstanding history of this subject.

Livio, Mario. Editor. *A Decade of Hubble Space Telescope Science.* New York: Cambridge University Press, 2003. An excellent recitation of what we have learned through the use of the Hubble Space Telescope.

McCurdy, Howard E. *Space and the American Imagination.* Washington, DC: Smithsonian Institution Press, 1997. A significant analysis of the relationship between popular culture and public policy.

McDougall, Walter A. *...The Heavens and the Earth: A Political History of the Space Age.* New York: Basic Books, 1985. Reprint edition, Baltimore, MD: Johns Hopkins University Press, 1997. This Pulitzer Prize-winning book analyzes the space race to the Moon in the 1960s. The author argues that Apollo prompted the space program to stress engineering over science, competition over cooperation, civilian over military management, and international prestige over practical applications.

Mather, John, and John Boslough. *The Very First Light: The True Inside Story of the Scientific Journey Back to the Dawn of the Universe.* New York: Basic Books, 1996. A solid account of NASA's Cosmic Background Explorer (COBE), written by the project's chief scientist.

Murray, Bruce C. *Journey into Space: The First Three Decades of Space Exploration.* New York: W.W. Norton and Co., 1989. This book offers an excellent discussion of the planetary science program written by the former director of the Jet Propulsion Laboratory.

Murray, Charles A., and Catherine Bly Cox. *Apollo: The Race to the Moon.* New York: Simon and Schuster, 1989. Perhaps the best general account of the lunar program, this history uses interviews and documents to reconstruct the stories of the people who participated in Apollo.

Neal, Valerie, Cathleen S. Lewis, and Frank H. Winter. *Spaceflight: A Smithsonian Guide.* New York: Macmillan, 1995. This book provides, with numerous illustrations, a basic history of space exploration by the United States.

——. Editor. *Where Next, Columbus? The Future of Space Exploration.* New York: Oxford University Press, 1994. An excellent collection of essays linking the voyage of discovery by Columbus with human exploration of space.

North, John. *The Norton History of Astronomy and Cosmology.* New York: W.W. Norton & Co., 1995. A lengthy scholarly account of the history of scientific understanding about the universe.

Reeves-Stevens, Judith, Garfield Reeves-Stevens, and Brian K. Muirhead. *Going to Mars: The Untold Story of Mars Pathfinder and NASA's Bold New Missions for the 21st Century.* New York: Pocket Books, 2003. An excellent account of the recent and future missions to the red planet.

Sagan, Carl. *Pale Blue Dot: A Vision of the Human Future in Space.* New York: Random House, 1994. Probably the most sophisticated articulation of the exploration imperative to appear since Wernher von Braun's work of the 1950s and 1960s.

Sheehan, William. *The Planet Mars: A History of Observation & Discovery.* Tucson: University of Arizona Press, 1996. An excellent survey of how humans have acquired knowledge about the red planet from antiquity to the present. It concentrates on the work of Earth-based astronomers but also includes succinct narratives of the *Mariner 4* mission and the Viking project of the 1970s.

Tucker, Wallace H., and Karen Tucker. *Revealing the Universe: The Making of the Chandra X-Ray Observatory.* New York: Harvard University Press, 2001. An important history of a major recent NASA space science effort.

Tyson, Neil de Grasse. *Universe: Down to Earth.* New York: Columbia University Press, 1995. An up-to-date analysis of the basic evolution of the universe using scientific information gleaned from research from all sources, including space exploration missions.

Zubrin, Robert, and Richard Wagner. *The Case for Mars: The Plan to Settle the Red Planet and Why.* New York: The Free Press, 1996. A critically important description of the "Mars Direct" strategy for reaching Mars with humans.

GLOSSARY

ANTI-MATTER Material composed of anti-particles, which are just like ordinary protons, electrons, and neutrons except they have opposite electrical charges and magnetic moments.

ASTEROID A boulder- to mountain-sized piece of rock remaining from the early solar system that moves around the Sun, mainly between the orbits of Mars and Jupiter.

ASTRONOMY The branch of physics that studies celestial bodies and all the matter-energy in the universe: its distribution, composition, physical states, movements, and evolution.

ATMOSPHERE The gaseous or air portion of the physical environment that encircles a planet. In the case of Earth, the atmosphere is held more or less near the surface by Earth's gravitational attraction.

THE BIG BANG The theory that the universe originated some thirteen billion years ago from the cataclysmic explosion of a small mass of matter at extremely high density and temperature.

BLACK HOLE An object with a gravitational field so strong that nothing, not even light, can escape from it. It is believed to be created in the collapse of a very massive star.

CONSTELLATION Groups of stars that seem to make up pictures on the celestial sphere. There are eighty-eight official constellations.

CORPUSCULAR UNIVERSE A theory that suggests that, at specific points in the universe such as black holes, there is a warping of space and time to create an additional universe that is parallel to our own. Cosmologists believe that there may be as many as eleven dimensions that relate to these multiple universes.

COSMIC RAY An electromagnetic ray of extremely high frequency and energy; cosmic rays usually interact with the atoms of the atmosphere before reaching the surface of Earth.

COSMOLOGY The astrophysical study of the history, structure, and dynamics of the Universe.

DARK ENERGY The residual energy in empty space that is causing the expansion of the universe to accelerate. Dark energy is spread almost uniformly throughout space and appears to contribute about 70 percent of the present energy density of the universe.

DARK MATTER A term used to describe matter in the universe that cannot be seen, but can be detected by its gravitational effects on other bodies. It is believed to make up more than 90 percent of the mass of the universe, but is not readily visible because it neither emits nor reflects electromagnetic radiation, such as light or radio signals. Its composition is unknown.

EXTRASOLAR PLANET A planetary body that orbits a star other than our Sun.

EXTRATERRESTRIAL LIFE Life beyond planet Earth (other than humans traveling in space, and living organisms they bring along or send).

GALAXY A large assemblage of stars (and sometimes interstellar gas and dust), typically containing millions to hundreds of billions of member stars. A galaxy is held together by the gravitational attraction of all its member stars (and other material) on one another.

GAS PLANETS A generic astronomical term invented by the science fiction writer James Blish to describe any large planet that is not composed mostly of rock or other solid matter.

GRAVITATIONAL FORCE The force of attraction that exists between all particles with mass in the universe.

INTERSTELLAR SPACE The space between solar systems that is made up of thin gas and very little else.

LIGHT-YEAR The distance light travels in a year (9.5 million million kilometers, or 5.9 million million miles).

MATTER The material (atoms and molecules) that constructs things on Earth and in the Universe.

METEORITE An object, usually a chunk or metal or rock, that survives entry through the atmosphere to reach Earth's surface. Meteors become meteorites if they reach the ground.

MICROGRAVITY A state where the gravity is reduced to almost negligible levels, such as during space flight or during free-fall.

MILKY WAY The spiral-shaped galaxy in which we live, made up of ten billion stars including our Sun.

MOON A secondary planet, or satellite, revolving about any member of the solar system.

NEBULA A cloud of gas and dust usually left by a dead star. Such clouds are where most stars form. They are primarily located by reflecting light from stars, emitting light from young stars that lie within, or by blocking the view of background stars behind them.

ORBIT The path of a planet, moon, or any other object in space as it revolves around another object, such as the sun.

PAYLOAD The cargo that is carried into space by a space vehicle.

PHYSICS The science of matter and energy and their interactions via measuring experiments.

PLANET Any of the celestial bodies (other than comets or satellites) that revolve around the Sun in the solar system.

SATELLITE An object that moves around another object. A satellite is kept in orbit by the equilibrium of gravity, which pulls it toward the object orbited, and centripetal force, which pulls it away from the object orbited.

SOLAR FLARE Brilliant flashes of light that develop suddenly in the Sun's atmosphere. Flares are usually associated with sunspots and are caused by the release of large amounts of magnetic energy.

SOLAR SYSTEM A group of large objects anywhere in space that has a star with orbiting planets. Our solar system is made up of the Sun and all things orbiting around it, including the nine major planets, their satellites, and all the asteroids and comets.

STAR A celestial body of hot gases that radiates energy derived from thermonuclear reactions in the interior.

SUNSPOT A small, cooler area on the Sun that shows up as a dark spot on it.

SUPERNOVA The death explosion of a massive star, resulting in a sharp increase in brightness followed by a gradual fading. At peak light output, supernova explosions can outshine a galaxy. The outer layers of the exploding star are blasted out in a radioactive cloud. This expanding cloud, visible long after the initial explosion fades from view, forms a supernova remnant.

TELESCOPE An instrument used to collect large amounts of light from faraway objects and increase their visibility to the naked eye. Telescopes can also enlarge objects that are relatively close to Earth.

UNIVERSE The sum of all existing matter, energy, and space.

WHITE DWARF A star, approximately the size of Earth, that has undergone gravitational collapse and is in the final stage of evolution for low-mass stars, beginning hot and white and ending cold and dark.

CREDITS

© Jeff Albertson/Corbis: 43A

© Paul Almasy/Corbis: 10–11

© Bettmann/Corbis: 66B

Wolfgang Brandner (JPL/IPAC), Eva K. Grebel (Univ. Washington), You-Hua Chu (Univ. Illinois Urbana-Champaign), and NASA: 65G

© William Coupon/Corbis: 58C

© Disney Enterprises, Inc.: 16C

ESA, Alfred Vidal-Madjar (Institut d'Astrophysique de Paris, CNRS, France), and NASA: 66A

John Frassanito & Associates: 50A, 50B, 50C, 51B, 51C

J. P. Harrington and K. J. Borkowski (University of Maryland), and NASA: 64A

Jeff Hester and Paul Scowen (Arizona State University) and NASA: 63B

Mohammad Heydari-Malayeri (Paris Observatory, France), and NASA/ESA: 64D

Hubble Heritage Team (AURA/STScI/NASA): 64C, 64F, 65A, 65B, 65D

John Jay/MPTV.net: 14–15A

Mars Society Image: 57A, 57B, 57C

Melies/The Kobal Collection: 14C

Courtesy Ron Miller: 14B

Moonrunner Design: 72

© B. Murton/Southampton Oceanography Centre/Photo Researchers, Inc.: 67B

NASA: 6–7 (AS10-32-4823), 16B (74-H-1056), 17, 18A (L-1959-5136), 18C (L-69519), 18D (74-H-1063), 22C, 22D, 23A, 23B (GPN-2000-1274), 24A, 24C (KSC-62-MA6-216), 26–27A (AS8-14-2383), 26–27B, 26A (64P-145), 26B, 27A, 27B, 27C, 27D, 28 (KSC-69PC-421), 29A (KSC-69P-631), 29B (KSC-69PC-413), 32–33A (ASI1-40-5903), 32B (ASI1-40-5877), 39C (KSC-89PC-469), 62–63 (STS082-746-59), 64O (STScI-PRC2002-14)

NASA, N. Benitez (JHU), T. Broadhurst (Racah Institute of Physics/The Hebrew University), H. Ford (JHU), M. Clampin (STScI), G. Hartig (STScI), G. Illingworth (UCO/Lick Observatory), the ACS Science Team, and ESA: 65O

NASA, ESA, and H. E. Bond (STScI): 65J

NASA, ESA, and J. Hester (ASU): 65N

NASA, ESA, and Mohammad Heydari-Malayeri (Observatoire de Paris, France): 64P

NASA, ESA, and D. Maoz (Tel-Aviv University and Columbia University): 64L

NASA, ESA, and Martino Romaniello (European Southern Observatory, Germany): 64Q

NASA, H. Ford (JHU), G. Illingworth (UCSC/LO), M. Clampin (STScI), G. Hartig (STScI), the ACS Science Team, and ESA: 64N

NASA/John Frassanito & Associates: 48–49, 51A, 56–57A, 56B, 56C

NASA, Andrew Fruchter, and the ERO Team [Sylvia Baggett (STScI), Richard Hook (ST-ECF), Zoltan Levay (STScI)]: 64C, 65M

NASA and The Hubble Heritage Team (AURA/STScI): 4–5, 8–9, 63A, 64E, 64H, 64J, 64I, 64K, 64R, 65E, 65H, 65I, 65K, 65L, 65M, 65P

NASA and Hubble Heritage Team (STScI): 64G

NASA/JPL: 22B, 38–39, 39A, 39B, 40A, 40B, 40C, 40D, 40E, 40F, 40G, 40H, 40I, 40J, 41A, 41B, 41C, 42A, 42B, 42C, 42D, 43B, 43C, 44A, 44B, 45A

NASA/JPL/Caltech: 20–21, 45B

NASA/JPL/Cornell: 45C, 45E, 46–47

NASA/JPL/U.S. Geological Survey: 45D

NASA/JSC: 24–25B (S65-63194), 32C (KSC-69PC-295), 53 (S83-35783), 67A (ASI2-48-7136)

NASA/MSFC: 52–53A, 52B, 52C, 52D

NASA image from the collection of J. L. Pickering: 18–19B

NASA/Pat Rawlings: 54B, 54D, 58A, 58B, 59

NASA/Pat Rawlings/SAIC: 54–55C, 54A

NASA, Donald Walter (South Carolina State University), Paul Scowen, and Brian Moore (Arizona State University): 65F

NASA/WMAP Science Team: 60

Novosti (London): 22A

Rare Book Division/Special Collections/J. Willard Marriott Library/University of Utah: 12, 13

Raghvendra Sahai and John Trauger (JPL), the WFPC2 science team, and NASA: 64B

© Dr. Seth Shostak/Photo Researchers, Inc.: 68, 69B

SETI Institute: 69C

SETI@home, University of California, Berkeley: 69A

University of Arizona Space Imagery Center: 30–31 (ASI1-40-5961, ASI1-40-5960, ASI1-40-5958, ASI1-40-5957, ASI1-40-5956), 34–35 (AS17-140-12494, AS17-140-12495, AS17-140-12496, AS17-140-12499)

U.S. Space & Rocket Center: 16–17A

TEHABI BOOKS

Tehabi Books developed, designed, and produced *Space: A Journey to Our Future* and has conceived and produced many award-winning books that are recognized for their strong literary and visual content. Tehabi works with national and international publishers, corporations, institutions, and nonprofit groups to identify, develop, and implement comprehensive publishing programs. Tehabi Books is located in San Diego, California. www.tehabi.com

President and Publisher: Chris Capen
Senior Vice President: Sam Lewis
Vice President and Creative Director: Karla Olson
Director, Corporate Publishing: Chris Brimble

Senior Art Director: John Baxter
Production Artist: Helga Benz

Proofreader: Lisa Wolff

Editor: Sarah Morgans

ISBN 1-931688-15-X

First Edition
Printed in Hong Kong
10 9 8 7 6 5 4 3 2 1